Lecture Notes in Mobility

Series Editor

Gereon Meyer, VDI/VDE Innovation und Technik GmbH, Berlin, Germany

More information about this series at http://www.springer.com/series/11573

Marin Marinov · Janene Piip

Editors

Sustainable Rail Transport

 Springer

Editors
Marin Marinov
Engineering Systems and Management
(ESM)
Aston University
Birmingham, UK

Janene Piip
JP Research & Consulting
Port Lincoln, SA, Australia

ISSN 2196-5544   ISSN 2196-5552   (electronic)
Lecture Notes in Mobility
ISBN 978-3-030-19521-2   ISBN 978-3-030-19519-9   (eBook)
https://doi.org/10.1007/978-3-030-19519-9

This Springer imprint is published by the registered company Springer Nature Switzerland AG
The registered company address is: Gewerbestrasse 11, 6330 Cham, Switzerland

Foreword

I am delighted to see that Volume 3 of *Sustainable Rail Transport* has been published. The fact that a third volume has been developed is itself a testament to the quality of the earlier volumes and the contribution that they made to the ongoing discourse in this hugely important field. My reading of the material in this volume suggests that it will become an indispensable work for experts in the field, as well as a valuable source of fascinating insights for non-expert rail enthusiasts like me. For rail specialists involved in both research and scholarship, as well as in professional practice, this volume provides a wealth of information and knowledge about key developments and trends impacting the sector.

The railway industry is a dynamic and challenging one. The market place has come more sophisticated with customers demanding ever-increasing value from providers. The wider business and regulatory environment is a complex one and subject to a high degree of change and unpredictability. Technology has continued to develop at a rapid rate, presenting rail professionals with an array of opportunities and threats—for many firms, simply keeping abreast of these developments is a major challenge. The anthropogenic impact of transport and logistics processes is now widely understood. In this context, the rail industry is well positioned to support wider policy initiatives aimed at creating more environmentally sustainable passenger and freight transport systems. Nonetheless, there is a pressing need to ensure that these systems are designed and managed in a way which minimises degradation to the natural environment whilst maintaining competitiveness in terms of cost and customer service. The myriad challenges currently facing the rail industry call for innovation in terms of both policy initiatives and operational practice. The publication of this volume is very timely in this context.

It is important that rail policy making is evidence-based and that innovation in the sector is based on the best available knowledge about processes, systems, technology and management. The high-quality research contained in this volume is critical in this regard. All of the work described is characterised by high levels of rigour in the research methodologies adopted by scholars. This is vital as the development of deeper and richer insights into the complex phenomena under investigation needs to be based on research designs that are logical and systematic.

However, this research excellence and academic rigour alone is not sufficient; it needs to be combined with a deep understanding of the evolving needs of the industry for the outputs to be truly valuable and impactful. I am delighted that this volume clearly demonstrates both academic excellence and practical relevance, with much of the work representing the fruits of effective academic/industry collaboration. The volume also demonstrates clearly that high-quality rail research requires truly interdisciplinary approaches. The implementation of the multi-paradigmatic philosophical positionalities required in this context calls for the effective adoption of methodological pluralism and the attendant use of mixed methods of data collection and analysis. None of this is straightforward, but the material in this volume provides many excellent examples of good practice in this regard.

Rich tomes of this kind require immense dedication and commitment on the part of a range of individuals—authors, editors, publishers and others. More importantly, it requires that the stakeholders work as a team with a clear view of the overall deliverable. In this context, the energy and enthusiasm of my Aston Logistics and Systems Institute colleague Dr. Marin Marinov warrants a particular mention. He is a passionate advocate of rail research and its critical role in developing a sustainable future for this strategically critical industry. The fruits of his endeavours, and those of his collaborators, are clear for all to see in this volume. *Sustainable Rail Transport* Volume 3 is a vital resource and can be read equally profitably by scholars and practitioners alike.

Birmingham, UK   Prof. Ed Sweeney
March 2019   Aston University

Preface

Sustainable Rail Transport Volume 3 will inspire rail industry academics and enthusiasts with new ideas and potential enhancements for current business practice. I have been honoured to receive Dr. Marin Marinov's invitation to co-edit this book and provide the preface for the subsequent chapters. Dr. Marinov is a dedicated advocate of sharing rail research and knowledge, having coordinated and co-coordinated all previous works in this series. He has passionately brought together various academics from different institutions to develop a global community interested in pursuing the dissemination of rail knowledge. In 2009, I became involved in rail industry workforce research, entering a fascinating career path. The opportunity to collaborate with other rail industry enthusiasts from around the world has taken me to more than 20 countries and provided countless topics to investigate.

Together, we are pleased to present this volume of innovative ideas from the contributing authors, showcasing insights from their recent research projects. Our previous two volumes of Sustainable Rail Transport were published in 2016 and 2017, receiving excellent feedback, so, once again, a group of dedicated academics and industry specialists have come together to contribute to this new edition. The authors geographically spread across the globe, in countries including the UK, Austria, Brazil, Egypt, Croatia, Thailand, Australia, Slovakia and the Czech Republic, have contributed chapters to make this a valued edition of Sustainable Rail Transport. Throughout the peer review and chapter development process, we have enjoyed reading the various versions of the author's work as the prepublication manuscripts were honed for the volume. The resulting material is impressively enlightening in terms of topics, coverage and innovation.

Today, because of the enormous advances made in the rail industry over the last two centuries, people around the world benefit from a mode of transport that encompasses both passenger and freight transport for their daily activities. With competition from other transport modes always at the front of mind of policy makers—from the road, maritime and aviation sectors—the rail industry is continually seeking new ways to keep ahead of the competition. The economics of rail transport means the industry is heavily reliant on economies of scale to provide services that are always going to be impacted by high fixed cost structures and relatively low variable costs. Innovation or sustainability is, therefore, a keyword that is important for the industry, as other modes of transport can always offer customer benefits such as cheaper and faster.

Some of the most attractive ways the industry can be seen of as sustainable, and a preferred transport option for both passenger and freight transport, is to disrupt the thinking of consumers in ways that meet customers' needs without customers having to change their behaviour. For example, the following chapters cover concepts around sustainability or 'the ability of a service to be maintained at a certain rate or level' such as 'clean and green', employer of choice, low cost and consideration of the environment.

The complex task of bringing the work together across countries and time zones is a testament to the dedication of the authors and editors. There has been a high degree of interest in contributing to the book, and I hope the readers will find the material useful as they explore more about the rail industry.

Port Lincoln, Australia   Dr. Janene Piip
March 2019

Contents

A Check-in and Bag Drop Service On-board Light Rail Vehicles
for Passengers Travelling to the Airport 1
Jonathan Toal and Marin Marinov

Quality Assessment of Regional Railway Passenger Transport 83
Borna Abramović and Denis Šipuš

The Possibilities of Increasing the Economic Efficiency of Regional
Rail Passenger Transport—A Case Study in Slovakia 97
Anna Dolinayova and Lenka Cerna

Sustainable Railway Solutions Using Goal Programming 129
Pedro Henrique Del Caro Daher, Diogo Furtado de Moura,
Gregório Coelho de Morais Neto, Marta Monteiro da Costa Cruz
and Patrícia Alcântara Cardoso

Parallel Genetic Algorithm and High Performance Computing
to Solve the Intercity Railway Alignment Optimization Problem 159
Cassiano A. Isler and João A. Widmer

The Use of Public Railway Transportation Network for Urban
Intermodal Logistics in Congested City Centres 187
Lino G. Marujo, Edgar E. Blanco, Daniel Oliveira Mota
and João Marcelo Leal Gomes Leite

Simulation of Fire Dynamics and Firefighting System for a Full-Scale
Passenger Rolling Stock 209
Ramy E. Shaltout and Mohamed A. Ismail

Novel Energy Harvesting Solutions for Powering Trackside Electronic
Equipment 229
Cristian Ulianov, Zdeněk Hadaš, Paul Hyde and Jan Smilek

High-Speed Overnight Trains—Potential Opportunities
and Customer Requirements 257
Bernhard Rüger and Peter Matausch

The Next Generation of Rail Talent: What Are They Looking
for in a Career? 275
Janene Piip

Professional Rail Freight and Logistics Training Programme:
A Case Study of Energy and Petrochemical Company in Bangkok,
Thailand 291
Kaushik Mysore, Mayurachat Watcharejyothin and Marin Marinov

Short Communication Paper 305

A Check-in and Bag Drop Service On-board Light Rail Vehicles for Passengers Travelling to the Airport

Jonathan Toal and Marin Marinov

Abstract It appears that nowadays rail vehicles are not the primary choice of transportation for people going to the airport. The inconvenience of carrying luggage on railways deters passengers, who look for alternatives. Attempts have been made to encourage passengers to travel to the airport by rail. However, significant limitations in these existing systems suggest a need for extensive work and adjustments, but this would increase the price and discourage passengers. This study investigates the potential for implementing an on-board check-in and bag drop system onto rail vehicles. By observing the Tyne and Wear Metro, Newcastle, the UK the benefits and limitations of installing such a facility have been explored, by the development of suitable operations and interior designs. Four designs which meet the design criteria were produced and their limitations considered. This study concludes that the potential for an on-board check-in and bag drop facility is realistic. Each design brings key benefits and limitations, and all meet security, and health and safety criteria. A feature incorporated into all designs allows for the equipment to be removed easily and stored away, helping with a low cost and versatile approach.

Keywords Metro · Rail vehicles · Check-in desk · Passengers · Baggage

1 Introduction

With the growing problems of increasing congestion on the roads and carbon emissions, new ways to travel are being investigated which are more efficient, environmentally friendly, and economical. Metro railway services are a viable solution to this problem. However, the majority of the public currently have a preference for

J. Toal
Mechanical and Systems Engineering School,
Newcastle University, Newcastle upon Tyne NE1 7RU, UK

M. Marinov (✉)
Engineering Systems and Management, Aston University,
Aston Triangle, Birmingham B4 7ET, UK
e-mail: m.marinov@aston.ac.uk

© Springer Nature Switzerland AG 2020
M. Marinov and J. Piip (eds.), *Sustainable Rail Transport*, Lecture Notes in Mobility,
https://doi.org/10.1007/978-3-030-19519-9_1

travelling by car [or road] (Department for Transport 2015). It is, therefore, the aim of companies to encourage people to use metro services more by offering a genuine incentive.

In busy cities, light rail can provide a reliable and direct service to airports. In 2015, the UK saw an estimated 250 million passengers arriving and departing the national airports per day (Civil Aviation Authority 2015). These figures highlight a crucial factor causing congestion on the roads. It is essential to address this issue by looking into new ways of transferring passengers. Most established airports are served by a rail facility. However, these rail facilities are not fully utilised, particularly by passengers carrying luggage. An extract from a recent presentation on the logistics of baggage less transportation quoted 'as long as the railway system is not able to replace the car boot, it will not be as successful' (Rüger 2016; Bernhard 2017). The difficulty of carrying luggage on public rail services deters people from using this form of public transport (GOV.UK 2011). It is common for passengers to opt for an alternative where baggage can be stowed away allowing the passengers to travel more easily. Investigating the potential for new rail facilities which offer improved incentives for passengers travelling in and out of the airport is crucial.

The Tyne and Wear Metro is a light rail service, operated by Nexus, which serves much of the population of the North East of England. There are over 70 stations, including one at Newcastle International Airport. The vehicles are poorly equipped for passengers with luggage travelling to the airport. Trains provide no storage for baggage with an internal layout which is challenging to navigate with luggage. These factors result in passengers opting for alternative modes of transport when travelling to the airport. As a result, the Metro is poorly utilised, customer satisfaction is low and a significant revenue generating opportunity is lost for the company (GOV.UK 2011; Wales and Marinov 2015; Darlton and Marinov 2015; Motraghi and Marinov 2012).

Using Tyne and Wear Metro as a case study, this research looks into solving the inconvenience of carrying luggage on rail vehicles by investigating the potential for implementing an on-board check-in and bag drop facility. It focusses on the interior product design needed to operate such a service on-board a rail vehicle, and the various designs and equipment that would be required as well as safety and security considerations. The external operations, technology and other facilities needed for a seamless rail service have also been evaluated and are introduced in this study. This facility is a new concept and is intended to develop a significant breakthrough in the understanding of whether this idea could be both beneficial and feasible.

Through the production of visual 3D digital models and investigations into the way the Tyne and Wear Metro vehicles would need to operate, the objective is to investigate whether a luggage drop off facility could be implemented on a light rail vehicle without major changes to the pre-existing construction. By considering the key benefits and limitations highlighted by this study, conclusions have been drawn about the potential of this new and innovative idea and, hence, whether the introduction of this facility could be feasible in Newcastle and elsewhere.

2 Motivation

In this current age, alternative options for transport are being explored to counteract the increasing production of carbon emissions through excessive use of petrol and diesel powered vehicles. In 2015, car traffic grew by a further 1.1% from 2014, reaching the highest levels since 2007 (Department for Transport 2016). This continuous increase in car traffic is a significant contributor to the high concentration of carbon emissions, particularly within city centres, while also creating significant congestion on the roads. Travelling by road is becoming unreliable due to the unpredictability of the journey and arrival times due to excessive traffic. A mode of transport which provides solutions to these issues is electric powered trains and metro vehicles. While road users account for 71% of transport carbon emissions; railway companies account for less than 1.8% (Henley 2013). Railway vehicles can transport several passengers per single journey during which the vehicle produces a negligible amount of carbon. This mode of transport removes vehicles from the road and contributes to the reduction of carbon emissions and road congestion. Particularly during peak times of travel, many commuters resort to travelling by rail out of choice. Rail provides an extremely reliable service, as the trains are efficiently organised to operate so that traffic is non-existent resulting in actual journey times that are more predictable. The service assures the passenger that expected destination times would be met when punctuality is vital.

Following the opening of the Tyne and Wear Metro in 1980, further extensions were made to the line. In 1991, the service was extended to Newcastle International Airport (Lewis 2012). The Tyne and Wear Metro now provides rail services for an expansive population in the North East of the United Kingdom, with stations located in Newcastle upon Tyne, Sunderland, North Shields and Newcastle International Airport. Despite this, full utilisation of this facility was not made. Since the introduction of the Tyne and Wear Metro operations, a decline in passenger flows has been observed with the metro scores being amongst the lowest in terms of operated passenger capacity in relation to network size (Nexus 2013; Motraghi and Marinov 2012). Increasing the number of passengers that use the Tyne and Wear Metro would have a positive impact on the requirement to reduce both carbon emissions and road congestion. One suitable way to do this would be to encourage passengers to use the metro when travelling to and from the airport. It is thought that while the current metro provides a service to Newcastle International Airport, the operations and interior design of the airport metro cars is not suitable or convenient for those passengers travelling with luggage. These factors deter people from using the service and instead, they resort to automotive transport.

The analysis of urban commodity flows is not new (Ogden 1977), though it seems that the passenger luggage has not been viewed as a commodity for urban freight (Woudsma 2001; Jaller et al. 2015). Urban freight by rail can work (Rüger 2016; Brice and Marinov 2015; Kelly and Marinov 2017; Dampier and Marinov 2015; Marinov et al. 2013; Motraghi and Marinov 2012) and this is a concept worth exploring as quite a few studies now confirm that freight transport in metropolitan areas is almost exclusively undertaken by trucks (Jaller et al. 2015; Dablanc 2007; Aditjandra et al. 2016). City logistics is an exciting concept, as we should stop considering each shipment, firm, and vehicle individually; instead, we should consider them as components of an integrated logistics system (Benjelloun and Crainic 2008), but does this concept include the passenger's luggage?

Due to the heavy luggage that a passenger cannot carry, travelling to the airport can be a stressful and expensive experience. For most people, the preferred mode of transport for airport transfer is automotive by use of car or taxi. These modes are particularly typical for those passengers travelling with luggage. The ability to travel hands free is the main reason for the preference to travel by car or taxi as large suitcases can be stored in the boot of the vehicle and handling of the luggage is kept to a minimum. These modes of transport can be costly. Travelling by car can result in paying for the extended stay car park facility provided by most airports. In the case of Newcastle International Airport, using such a facility can cost more than £80 per week. Travelling by taxi can also be associated with extremely high costs, particularly for those that live further away from the airport. In the case of someone living in Sunderland, a return taxi journey to Newcastle International Airport can exceed £100. It is for those passengers wishing to travel to the airport but wanting to avoid these high transfer costs that a more affordable but still convenient option may appeal. The Tyne and Wear Metro can provide a suitable mode of transport for people travelling to the airport as it is both cheap and reliable. Despite this, the use of the metro is avoided due to the inconvenience of having to carry large luggage on a public transport service. Travelling to Newcastle International Airport by road can be unreliable due to excessive congestion, particularly during peak times of rush hour. Although the metro avoids this issue, overcrowded metro carriages is another factor that discourages passengers from using this facility as boarding the metro with baggage is currently impractical, particularly in the case of a family with several suitcases.

3 Existing Systems

3.1 Virgin Bag Magic

Virgin Bag Magic is a service introduced by Virgin Trains. From £31.67, a courier can collect a passenger's suitcase from their house and deliver it to any UK destination. This service removes the need for passengers to carry a heavy suitcase with them and provides for more pleasant travel. Knowing that the train journey can be made without luggage; that they will not have to struggle to store it or worry about leaving it out of sight, provides peace of mind and reduces concerns for the customer. By using this facility, passengers can leave their house, travel by public transport and arrive at their destination without their suitcase. This service offers the added benefit for more opportunities, such as, upon arrival, heading straight to a business meeting or restaurant without the need to drop off any luggage at the hotel beforehand. Such a system could be adapted to the Tyne and Wear Metro to encourage passengers to use the metro service by removing the issue of passengers being hindered by their luggage.

While this service seems an adequate solution to the issues raised, it does have some substantial drawbacks. In the case of people travelling unexpectedly at the last minute, the Virgin service would not be available as suitcases are required to be collected from the person's home one day in advance. There is a disutility associated with passengers having to pack their luggage early; consequently, this makes this service unsuitable. Another limitation of this service is that it cannot be used when travelling abroad. Passengers who are flying to a destination outside of the UK are not able to access this service and is currently extremely limited in terms of its global availability. Finally, the most critical restriction with this facility is the cost. With this provision costing over £30, in some instances, the fee matches that of the train ticket. This service results in a 100% increase in the cost of the journey rendering the Bag Magic facility less of a prudent choice. In Reece and Marinov (2015a), Virgin Bag Magic was seen as a facility which would be better utilised providing the costing was reduced or even free.

3.2   InPost

InPost is a service in which people can send a parcel and have it delivered to one of many national, global lockers positioned across Europe. A postage stamp can be printed online and attached to the parcel before either being handed to the courier or by placing the parcel in one of the available lockers. The parcel is then transported to the required destination in which it is then stored in a locker, ready for the recipient to collect. The use of this facility can be related to that of a baggage handling system. Passengers travelling to destinations abroad can use the facility to post items to a location near to their hotel to remove or reduce the number of items they are to carry, allowing them to travel light (InPost 2016).

The InPost service works similar to that of Virgin Bag Magic with the added benefit of the delivered item being able to be collected at an InPost locker at any time. Issues arise in that the maximum dimensions for the parcel are limited at 38 cm × 38 cm × 60 cm which equates to a small suitcase. Such a facility cannot be used for large heavy luggage which is the main issue when it comes to the impracticality of carrying luggage on public transport. Furthermore, when posting abroad, the minimum number of days to deliver is on average six days which would require advance preparation in sending an item. As stated in Reece and Marinov (2015a) of Virgin Bag Magic, such a system which requires early packaging is not advocated by most passengers and, hence, the InPost service would be highly impractical.

3.3 Hong Kong in Town Check-in

Several baggage handling services have been erected in different cities across the world in an attempt to encourage people to use public transport when travelling to the airport. An example of a system like this is Hong Kong in town check-in. Passengers can check in their luggage in the city centre before then being able to travel hands free for the remainder of the day leading up to the departure of the flight. The checked in luggage is then transported as freight via a separate train which carries the luggage to the airport to be loaded onto the aircraft. Passengers can then spend the day exploring the city and then continue to the airport without luggage and proceed straight through to security thus avoiding excessive queues to check in bags. This facility offers a great incentive to use public transport. It minimises the time required by the passenger to be at the airport before flight departure while also allowing passengers the freedom to travel without baggage. Although people, particularly those with higher incomes, may have insisted on using a taxi or car to travel to the airport, they are now provided with a viable option to use public transport that provides the benefits that taxis and cars cannot offer. By encouraging more people to travel by public transport, the stress of being stuck in traffic against time constraints and removing the costly fees for parking and taxi fares. In addition to the physically demanding task of manoeuvring bags through turnstiles, up and down stairs, and between crowds of fellow passengers, is eliminated (Jaffe 2014). Furthermore, this facility costs just $13 U.S which is cheaper than other forms of transport and, to many, may seem financially viable.

An initial downfall with this facility is the issue that passengers are required to travel to the city centre with their luggage, necessitating passengers to travel, for part of the journey, with a form of luggage. The inconvenience of carrying a suitcase on public transport has not been entirely removed. There is also the potential for extended travel time in that passengers would be required to depart via the city centre public transport service in order to check in their bags. The reality is that it may have been quicker and more direct to travel straight to the airport. For people living closer to the airport than the city centre station, there exists the possibility that they would have to travel away from the airport and into the city centre in order to first check in their bags. These processes are counterproductive and would add more time to the journey.

3.4 Baggage Collection Hub at Haymarket Station

Brice and Marinov (2015) conducted a study in which a baggage collection system was modelled to observe whether such a system, similar to that of the Hong Kong in town check-in facility, could be implemented with the Tyne and Wear Metro. The idea was that a baggage collection hub would be located in the Haymarket metro station (Reece and Marinov 2015a, b) in which passengers could check in and drop off their luggage before proceeding to Newcastle International Airport.

When analysing this model, an estimate of just 24 passengers per year travelling from North Shields were calculated in using this facility out of a current metro ridership of 6000 (Reece and Marinov 2015b). This value is extremely low and shows a much lower utilisation of the facility than that estimated for Virgin Bag Magic or the InPost service. The main reason for this low estimation of passengers was found to be due to the need for passengers travelling from North Shields to transfer the metro line at Monument station and then depart at the following station (Haymarket station) in order to check in their luggage. The need for multiple alights is incredibly inconvenient for passengers; it adds to the journey time; it is awkward with heavy and multiple baggage and is not a relaxing way to travel. It would appear that direct travel on the metro to the airport station, and being able to avoid multiple manoeuvres of luggage on and off the vehicle, would lend itself better to customer convenience. Cost is another factor highlighted that contributed to the low utilisation of the facility by the passengers. The study found that when setting the cost to zero, the number of passengers that would use the baggage collection hub out of the 6000 metro riders increased by 113.4% (Reece and Marinov 2015b), showing a demand for such a system. It would require consideration of low costs to the user, or deemed as psychologically free, by including the price within the flight or metro ticket. This challenge may be difficult as such a facility would be extremely expensive for rail companies or governments to provide, with major works required on the infrastructure of the metro at both Haymarket and the Airport station. Changes to the current signalling would have to take place, and several new employees would be required. All these factors contribute to the need for a significant financial investment. With a small number of passengers estimated to use this facility, the price to access such a system would have to be considerably high in order to fund this concept.

3.5 Baggage Transfer System Using Tyne and Wear Metro

The potential of the Tyne and Wear Metro being used to transfer baggage by running a pendulum freight train system between the Haymarket metro station and Newcastle International Airport was investigated by event-based simulation in a study that conducted by Brice and Marinov (2015).

The results of this study identified that the metro vehicles could be used with a capacity of 9750 bags across 26 freight train journeys with potential for the metro to transfer 61,125 bags per day. It was concluded that the metro had the potential to be enabled to transfer freight. This study supports the opportunity for the metro to carry large quantities of luggage with an on-board bag drop facility. Luggage could be stored on the metro and developed into a working and successful facility.

Fig. 1 Proposal for the interior design for the Melbourne Airport Monorail (Monorails Australia 2014)

3.6 Melbourne Airport Monorail Interior Design

While new external facilities which help remove any luggage from a passenger is beneficial in encouraging the use of public transport services, having an interior design for the metro which is suitable for accommodating passengers carrying large bags is also fundamental. Currently, the Tyne and Wear Metro is inappropriately designed for passengers travelling to the airport with luggage. The metro contains no storage on board and has narrow walkways, which make carrying luggage through the metro a daunting task. This in itself could be a significant factor that contributes to why passengers prefer to travel to the airport by taxi or car. To further progress a baggage handling and storage concept, the interior design of the metro must also be addressed.

An example where the interior design has been considered is in the proposal for a high-capacity automated monorail to Melbourne Airport, Australia (Monorails Australia 2014). The design produced for the internal layout of the monorail has considered the high capacity of passengers with baggage, as seen in Fig. 1. The use of longitudinal seats has been integrated into the design. This provides open space and makes it easier to manoeuvre luggage through the vehicle. It creates space in front of seated passengers, making it possible to place any luggage by their feet. Hanging straps and vertical rails have also been used to provide safety for passengers who are standing. This internal design would encourage people to use this rail service as a means of travelling to the airport. The design features should be contemplated and carried forward into the Tyne and Wear Metro.

If considered in isolation, upgrading the internal design may not be sufficient to see a significant increase in the utilisation of the Tyne and Wear Metro, when travelling to Newcastle International Airport. There remains the inconvenience of having to handle and maneuverer a suitcase on public transport.

The design should consider combining a baggage collection facility with an upgraded internal design within the metro.

3.7 Tyne and Wear Metro

The metro is made up of two services called the Yellow Line and the Green Line. The Yellow Line operates a service from St. James to South Shield Station while the Green Line operates from South Hylton to the Airport Station, as seen in Fig. 2. For this study, the Green Line service will be observed. A metro vehicle leaves from South Hylton every 12-15 min depending on the time of day (Nexus 2014). Nexus stated the turnaround time at both South Hylton and the Airport Station should not exceed 10 min.

The main factors that influence the demand for the metro are the population, GVA (Gross Value Added), road fuel costs and impacts on congestion. Currently, the North East has a growing population with an increase of 2.2% from 2001 (Office for National Statistics 2012). The GVA had seen a massive 90.8% population increase in the North East since 1997 (Harari 2016) while petrol had gone from 28.2p per litre since 1980 (when the metro operations began) to over 110p per litre (Nexus 2013). Congestion is increasingly becoming an issue, particularly in the city centre. Based on these facts, there should be an increasing demand for the metro. However, the utilisation of the metro is not maximised with the potential for increased ridership. Further incentives should be made to encourage the use of the Tyne and Wear Metro which can be done by installing facilities for passengers travelling to the airport to accommodate those carrying luggage.

In the document, Metro Strategy for 2030 (Nexus 2013), it states that there is continued growth in passenger numbers forecasted by the airport operator with the number of passengers travelling to the airport is the highest growth in percentage

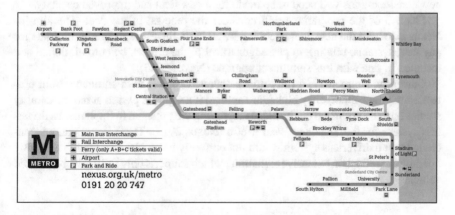

Fig. 2 Map of Tyne and Wear Metro Rail Service (Metro et al. 2016)

terms. This passenger growth supports the concept that there is a necessity to consider the upgrade required to accommodate better those travelling to the airport. It is clear that people are already willing to use the service to travel to the airport; therefore incorporating more facilities to incentivise passengers travelling with baggage would significantly increase the ridership of the Airport Metro.

Compared to other light rail services within Europe, the Tyne and Wear Metro has one of the highest seating densities. These facts do not reflect that in Nexus (2013), most passengers travel on the metro for a total of 10 min and do not necessarily require a seat, demonstrates the potential to remove seats within the carriages to better accommodate passengers travelling to the airport with luggage. In the case of the Edinburgh Tram, the seat density was dropped to just 1.8 (compared to the Tyne and Wear Metro's of 2.3) to accommodate for storage space required for Airport traffic (Nexus 2013). This example is something which should be considered for the Tyne and Wear Metro as doing so would encourage airport passengers to use the metro service while also offering an increased capacity.

By 2025, Nexus is aiming to have produced significant upgrades and modifications to the existing metro rail service currently in place. The regeneration process for the metro has been called 'The Metro Reinvigoration' programme, sectioned into three phases.

Phase one has been completed, involving upgrades of the ticketing and gating programme surrounding the metro. Meanwhile, Phase two is currently in process and aims to complete modifications to the track, building, systems and stations. Refurbishment of the tracks focusses around the city centre, the refurbishment of 45 stations is in the process, and upgrades to the communications and signalling systems are yet to be conducted. As this work has already commenced, it may not be viable to look at installing baggage collection schemes at stations in Newcastle city centre. Such a system would require further upgrades to the infrastructure and signalling which would be difficult to fund at this late stage. Within Phase two a significant condition assessment has been conducted on all 90 of the metro cars which determine them suitable for continued operation into the 2020s. By this time, new considerations will need to be made for a new fleet and improved capacity.

Phase 3 of the programme will consider the renewal of the fleet. The aim for Nexus is to have the new fleet operating by 2025. It, therefore, seems appropriate for any new systems, relating to passenger travel to the airport, be commissioned and installed in line with this significant upgrade (Nexus 2013).

Currently, the metro is a fully driver-operated vehicle with automatic train protection in which emergency brakes are applied if a signal is passed at red. Feasibility studies have been carried out by Nexus regarding the potential to implement a driverless system for the Tyne and Wear Metro (Nexus 2013). These investigations concluded that a driverless system would not currently be viable. The facility currently operating is reliant on extensive signalling, which supports only fully driver-operated vehicles.

3.8 Newcastle International Airport

Newcastle International Airport is the largest airport in the North East of England. It operates out of one terminal building, servicing 20 different airlines. Since 2010, the number of passengers who utilise the airport has grown from 4.34 million to 4.54 million per year (Passenger Statistics 2016). This statistic shows there can be as many as 12,500 passengers per day using the airport with high volumes of road traffic and the need for an incentive for these passengers to use the metro service. A need to support the adaptation of the metro to accommodate passengers travelling to and from the airport would significantly increase the ridership and generate more money towards the facility. The increasing number of passengers shows the need for change to the metro to facilitate better transportation for these passengers.

The metro station located at Newcastle International Airport provides stair-free access into the airport terminal thus making transporting luggage from the platform a simple task. The metro station is also equipped with a sheltered roof top which can help protect luggage from adverse weather conditions. As demonstrated, some facilities to implement these ideas are already in place.

4 New Service

Through the review of the baggage handling systems already in existence and the analysis of the data collected from studies, we considered a new service that could apply to the Tyne and Wear Metro. This proposal is believed to solve the issue of poor public transport utilisation when travelling to the airport while addressing the limitations of the existing solutions discussed. The solution is to investigate the potential of implementing an on-board check-in and bag drop facility for passengers travelling with luggage to the airport. How a light rail service would be required to operate to facilitate this new concept will be explicitly stated for this study while the necessary equipment and the functionality and layout of the interior of the railway vehicle will be assessed through the study of the current Tyne and Wear Metro vehicles and operation. Such a facility would allow passengers to check in their luggage upon boarding and therefore enjoy the majority of the journey without any baggage. This study will specifically look at the interior product design of the on-board check-in and bag drop facility and investigate how such a facility could be retrofitted into the existing Tyne and Wear metro cars using a variety of unique designs displayed as visual 3D digital models.

This new idea offers many incentives that other modes of transport and existing facilities cannot provide. A great deal of time is saved because the time taken to travel to the airport and check-in luggage is combined into one journey hence making the time required for the standard procedures before the departure of the flight minimal. Any anxieties of excessive queues at the check-in desks inside the airport terminal are also eliminated as passengers who use the on-board check-in and bag drop facility can proceed straight through to security upon arrival and so offering an appealing incentive to passengers of all incomes.

Significant regeneration to the infrastructure and stations is avoided by locating the check-in and bag drop facilities inside the metro vehicle. Initial scoping suggest this option to be low cost in terms of installation. The service offered to passengers could be competitively priced to make it a financially viable option for customers. Furthermore, the on-board check-in service would allow passengers to immediately check in their baggage upon boarding the metro vehicle, removing the need to depart mid-way to check in luggage. In the case of bag drops located at one specific station, this service would allow passengers to travel without luggage on their person. This study aims to test the true capabilities of this facility and assess whether these initial beneficial factors could be viable options that could be put into practice depending on the discoveries made through the development of this study.

5 Methodology

The identification of the requirements, characteristics, market research and facilities available for this study was conducted using a road mapping technique. This methodology provides a framework to help tackle the fundamental questions that apply to this study (Univeristy of Cambridge 2010) and clearly demonstrates its aims.

By observing the design and operations of the current Tyne and Wear Metro vehicles through both desktop research and first hand visits to Nexus metro depot, we developed several design solutions for the interior of the Tyne and Wear Metro cars. We also explored existing facilities and technology to inform our design using both new and existing equipment and alternative methods of operation, to facilitate the on-board check-in and bag drop system.

Various equipment, operations and interior layouts were considered before producing a selection of sketches displaying alternative designs. Following evaluation of these designs, we modelled and analysed a selection of final designs fulfilling the set criteria.

The final designs have been created as a 3D digital model using Autodesk Inventor to give a visual display of the unique on-board check-in and bag drop systems. An accurate original 3D model of the existing metro vehicle design was produced before the new system designs were implemented into this model. These 3D digital models give a clear and realistic picture of the setup of the facility on the metro, enabling a reliable visualisation which offers assurance in the functionality of the design idea.

Through using design evaluation techniques and public opinion of the designs, the potential of an on-board check-in and bag drop facility provided insight into the benefits and limitations of this baggage handling and storage concept. Furthermore, mechanical validation showed that the metro could facilitate the designs in terms of their functionality, layout, size and weight. This outcome provided further evidence that such a facility could be installed on the metro vehicles in Newcastle.

6 New Service Development

6.1 Design Criteria

In order to produce a set of suitable solutions as to how an on-board check-in and bag drop facility could successfully function, key criteria were acquired through a literature review and experts opinion. The design criteria had a clear objective regarding the essential factors in the development of design solutions. By taking the sum of the scores, a percentage weight for each design criteria could be produced based on the importance given for each criterion and final evaluations of the products. Table 1 shows the key criteria and their importance.

Performance (5)—It is vital for the designs to provide maximum performance in the facility provided including optimising the volume of luggage the designs can accommodate, providing a wide range of facilities for passengers to utilise and operating efficiently.

Economical (5)—The solutions produced must have a heavy weighting for the costs required to implement an on-board check-in and bag drop facility for rail vehicles. Designs should be realistic about the technology and equipment required while the more economical options should be favourably considered. The lower the cost of installing this facility onto rail vehicles, the cheaper the service can be provided to the customers, subsequently attracting more users of the service and generating higher revenue for operators.

Versatile (5)—Cognisant of the current Metro Strategy, the designs must be developed in conjunction with the current infrastructure, systems and vehicles. By developing solutions which can be adapted to the existing operations and vehicle designs, a more valid, assured and cheaper design will be produced. The designs should attempt to minimise the impact on the current capacity and layout of the metro vehicle with the least disruption to passengers and operations, ensuring customer acceptance of this new facility while allowing the metro to operate efficiently as usual. It is vital that as many customers as feasible can be comfortably accommodated within the

Table 1 A table showing the design criterion set for this study

Design criterion	Importance (1 low, 5 high)	Weight (%)
Performance	5	14.7
Economical	5	14.7
Versatile	5	14.7
Safe	5	14.7
Secure	5	14.7
User friendly	4	11.8
Robust	3	8.8
Aesthetically pleasing	2	5.9

carriages. Improved customer satisfaction will lead to an increase in public interest and further uptake of this service.

Safe (5)—Alongside any new design, the safety of staff and customers are paramount if a design is to be accepted and developed. Ideas which create uncertainty surrounding the safety of any person must not be considered further.

Secure (5)—Security of the luggage is a necessity in the solutions designed for this study. Any doubt inferred from possibilities of the luggage being tampered, lost or stolen will deter passengers from using the check-in facility. While extensive security checks may not be possible in the solutions produced; alternative options should be considered and attempted to aid in the assurance of both luggage and thus passenger safety.

User Friendly (4)—In terms of the customers, using this facility should be equally as straightforward as checking in luggage at an airport. The idea is to attract passengers by offering a stress free service with beneficial attributes of a car or taxi. Consideration must also go into the ease of use of the equipment for the members of staff to retain employee satisfaction and ensure high productivity, excellent customer service and efficiency in the operation of the equipment. Furthermore, it is essential that staff members are not required to carry out overly demanding tasks which could result in injury. Health and Safety Regulations must be considered. Where economic viability, versatility, safety or security factors are significantly impaired due to user friendly functionality, priority should always be given to these more critical criterions.

Robust (3)—Equipment which is heavy or transported frequently, must be able to withstand different environments and have the robustness to be able to withstand knocks, outdoor conditions during transportation to the metro vehicles and heavy loads. A complete assessment of the robustness of such equipment cannot be completed until a prototype of the proposed solutions can be manufactured.

Aesthetically Pleasing (2)—Equipment used, and the setup of the facility should be visually appealing to customers as it reflects a user friendly and quality service which initially attracts customers. The main focus of this study is on looking at the potential for a check-in and bag drop facility to function on-board a rail vehicle. Slight alterations to the presentation of the designs could be made before a prototype being produced. However, it is the functionality of this system which takes priority in the study.

6.2 Luggage Sizes

In the case of this study, the dimensions in Table 2 have been used to represent the various luggage sizes which are to be considered for the storage design. How each size of luggage is represented in the 3D model designs is displayed under 'Model Presentation'.

Table 2 Dimensions for luggage size

Size	Height (m)	Width (m)	Depth (m)	Volume (m^3)	Model presentation
Large	0.76	0.48	0.29	0.11	
Medium	0.67	0.45	0.25	0.08	
Small	0.63	0.36	0.21	0.05	

6.3 Location of Check-in Desk

Two areas on the metro were initially deemed as being suitable to install a check-in desk. These areas were at location 1, approximately a quarter of the way from the front of the metro car, and location 2, at the centre of the car. It should be noted that in the case of the Tyne and Wear Metro, each metro car is numbered from one to ninety whilst the ends of the cars are labelled as A and B. Figure 3 displays the current seating arrangement for the Tyne and Wear Metro and highlights the locations considered for the check-in desk to be erected whilst also labelling end A and B for this particular drawing of the car.

Location 1 was initially analysed due to the low seat density to the front of the metro. The front end of the metro was observed as having more open space relative to the rest of the metro, being ideal in terms of being able to store checked in luggage and minimise the need to remove a vast number of seats. In order for this idea to work, the doors to the front of the metro had to remain closed to accommodate the luggage for the duration of the journey until arriving at Newcastle International Airport. Airport passengers intending to utilise the check-in desk would then be required to board the metro through the second set of doors from side A. From here, these passengers would move through the metro and towards location 1 where luggage could be checked in and removed from the passenger. The passenger would move then move back up through the metro, baggage free, and be seated with standard passengers at side B of the metro while the luggage is stored at the front end of the metro.

Fig. 3 A caption of the current seating arrangement of the Tyne and Wear Metro which highlights location 1 and 2

Fig. 4 Centre area of the metro vehicle where location 2 check-in desk would be installed

Location 2 was considered as only longitudinal seats are installed at this point, creating an area which offered good floor space and seating, with a large surface area suitable for stacking luggage. This concept can be seen in Fig. 4. For this location, it was proposed that airport passengers would board the metro through the same set of doors as with location 1 with passengers moving through to the centre of the metro. Following luggage check-in, passengers would then either be seated in the available longitudinal seats within the centre of the metro or move through to the standard seating area located at side A of the metro.

Observing the interior of the metro vehicles we compared, two potential locations for the check-in desk and uncovered several factors which influenced the optimum location for the check-in desk. In the case of location 1, passengers are required to move up and down between the narrow aisles located in front of the check-in desk. This process can be very impractical particularly when moving towards the check-in desk and manoeuvring luggage between the wide, obstructive seats possibly having a deterring and negative impact on passengers who are considering using the metro when travelling to the airport. Furthermore, congestion on-board the metro from passengers moving up and down the narrow aisles, could cause overcrowding, safety issues and impede the operating doors. Within location 1, there is also little room to install a check-in desk that sits adjacently to the side walls of the metro as the presence of a narrow aisle remains. Due to these factors, all the seats from location 1 to the second set of doors, on side A of the metro, would have to be removed, re-designed or repositioned to create open floor space. The ability to remove these seats is not feasible as most contain equipment beneath them, such as saloon heaters, a PTI supply and a door isolator. It would be necessary to either relocate this equipment or redesign the layout of the seats so that they are repositioned as longitudinal seats, a feature already installed in the centre of the metro. Both of these options are costly and time consuming, particularly when it is considered that this would have to be applied to most of the fleet of vehicles. A permanent change to the seating on the metro would limit the potential for the metro to operate as usual when the check-in and bag drop facility is not in use. It would create dissatisfaction for regular passengers who use the metro for reasons other than travelling to the airport. It seems sensible in this instance to locate the check-in facility at location 2 where the seating is already designed. It has open floor space and no change to the positioning of the seats is required.

Recently, perch seats were installed at either end of the metro to accommodate the RVAR regulations for accessibility of disabled passengers with wheelchairs to ensure the chairs can board the metro vehicles safely. The position of this wheelchair area has explicitly been located to meet requirements for the LED metro information board to be visible from either direction (RVAR 2015). Positioning the check-in desk at location 1 obstructs the wheelchair area and does not meet RVAR regulations, creating another undesirable additional cost and lengthy procedure in relocating a facility which has recently been refurbished. Converting the centre of the metro into a check-in area would not influence the pre-existing RVAR required facilities and therefore provide a cheaper and simpler instalment of a check-in and bag drop facility.

Nexus must follow the health and safety requirement which states that there must be at least six operational doors per metro car at any given point. While the plan of operation, with the check-in desk positioned in location 1, does allow for six operational doors to be functioning at once, it does create limited resilience for the vehicle if one door were to fail during transport. If this were to occur, then the service would have to cease operating immediately and be transported to the Nexus depot workshop in order for it to be repaired. In the event of this happening, significant complications can occur about the logistics of luggage already checked in on the metro and ensuring passengers do not miss a flight. The need to prematurely remove luggage from the metro before arriving at the airport could create major security concerns and complications in ensuring all the luggage is safely transferred to the airport. By installing the check-in desk in location 2 of the metro, all doors can operate as usual as no part of the check-in facility would overlap any sets of doors. Not only does this not reduce the existing reliability of the metro but it also further prevents the need to apply any mechanical change to the metro car. It is important not to hinder any existing positive attributes about the metro such as its reliability, mainly when this is already a key benefit over automotive travel.

Storing luggage at the front of the metro (side A) can cause complications in terms of access and egress for the driver to the driver's cabin. A large amount of space at the front of the metro could not be utilised for storage as it would prevent the driver opening the cabin door thus the space available for luggage at the front of the metro is more limited than initially assumed. Health and safety concerns are considered in the event of an accident in which luggage could pile up against the driver's cabin and prevent the driver's quick escape.

The centre of the metro is supported by a tractor bogie with the centre bogie pulled along by an engine at the front of the metro and no added weight in terms of an engine and gearbox. The ability to add new equipment and facilities in this area with the potential for storing large amounts of heavy luggage is further assured as this is the best location to have added weight introduced.

The conclusion of this analysis, to identify the optimal place to locate the check-in desk, suggests that location 2 provides a more prefabricated area with more significant potential to be adapted to operate as a check-in and bag drop facility without the need for extensive refurbishment. In turn, this concept allows for the possibility of a more cost effective upgrade to the metro with a more versatile area. For these reasons, location 2 is considered as the area to be investigated for this study.

6.4 Operation of the Metro

With the number of flights departing Newcastle Airport each day to various destinations, it can become a complex system in which high volumes of luggage bound for several different destinations need to be checked in, from one check-in desk onboard the metro, and then accurately directed to the relevant airline. If passengers can board and check-in luggage on any desired metro during that day, it can become overwhelming for staff. Checking in passengers and printing the correct baggage labels corresponding to the destination of the luggage from one desk, while also making the number of passengers expected on each metro vehicle unpredictable. Attempting to facilitate the check-in desk in this way can override the system, create longer waits to check-in luggage, lead to higher costs and complex organisation, and therefore this type of operation should be avoided.

Following luggage being checked in at an airport, the baggage is transferred by a conveyer belt, passed through security checks and finally sorted and loaded to different flights (Savrasovsa et al. 2007). The metro check-in service must follow a similar schedule to that used by the airport to ensure that luggage can be sorted to the correct destination along with the non-metro baggage, checked in as usual at the airport. Communication between the metro and the airport is critical in this facility being a success. To assist in these factors, we concluded that for a check-in and bag drop service to operate on-board a metro, it would be a more sensible assumption for each metro to operate for different airline companies with the check in for particular flights correlating to departure times at the airport. This arrangement allows luggage to be transferred from the metro at designated times to the corresponding airline. It allows the computer used on the check-in desk to be programmed before the departure of the metro from the first stop (South Hylton). Through the use of a clear communication strategy between both Nexus and Newcastle International Airport, luggage can be collected from the metro at the Airport station with a prior knowledge of the airline, destination and sum of luggage expected, thus ensuring exceptional levels of organisation.

For each day, a planned schedule can be produced to allocate metro cars to specific airlines and flights, depending on the flight departures that day. Not only would this organisation of the rail vehicles aid in preparing the correct equipment required on-board that specific metro, but would be understood by Nexus operators what airline and destination of luggage to expect. Extensive surges of passengers during peak times would be overcome as the maximum passenger expectancy can be calculated previously and managed accordingly. During off-peak times, the on-board check-in and bag drop facility can serve a broader range of flights over a broader time scale. Where there is a high density of flights set to depart, or large passenger airlines are in operation, then the range of destinations and the check-in desk on-board the metro serves can be limited to control the passenger flow.

It would not be necessary to have the entire airport metro fleet operating the on-board check-in and bag drop system mainly if an effective schedule of the metros can be accomplished. Currently the Tyne and Wear Metro provides a service in which a vehicle departs for the airport on average every 15 min (Nexus 2014). By confining the check-in availability on each metro relative to the airline company and the time in which flights are due to depart, the number of passengers can be controlled, and utilisation of each metro and frequency of services can be adjusted. Nexus could provide this service with every other metro thus offering it twice an hour, saving money for equipment purchase and reducing staffing costs.

6.5　Customer Facilities

With a new baggage system, there must be facilities available to educate passengers on how the system operates. An obvious method is to provide information through the Nexus website relating to the capability of the facility, how tickets can be bought, where passengers need to board and journey information. A weekly schedule, posted online would outline times for metro vehicles with on board check-in facilities and the flights served by the metro vehicles. Passengers could input their flight details into a search engine on the Nexus website to locate the required information. The website functions could be incorporated into a smartphone app for live updates. Real-time departure information would enhance the customer experience (Tyne and Wear Public Transport Users Group 2010). Passengers should have access to a help line phone number or flyers and poster which can be sited at most stations on the airport line as well.

Within the stations on the airport line, information must be provided to ensure clarity to passengers, such as the on-board check-in and bag drop facility and where they must board the metro. Intelligible signs placed around the platform and markings on the floor where airport passengers should position themselves before the metro arrives will be necessary. To help further clarify the correct location where passengers are required to board, bold text and images should be displayed on the outside of the doors offering instructions to the passengers.

The front of the metro should display flight information on the existing LED monitor to offer assurance for passengers in boarding the correct metro relative to the flight. Public announcements should be made at the platform ahead of the metro arriving also confirming the flights which the incoming metro serves.

6.6 Initial Design Limitations

During research, several limitations were discovered which were influential in the designs, and these were overcome to ensure the production of realistic and functional solutions.

The Green Line operates to the Newcastle International Airport station. There are issues installing a driverless system due to the presence of pedestrian level crossings and the metro sharing the same line as Network Rail's Sunderland Line (Nexus 2013). The current signalling and infrastructure of the Tyne and Wear Metro is not equipped to facilitate a driverless service and would require significant changes. While installation of a driverless system has the potential to be incorporated in the future, such a modification to the Tyne and Wear Metro would be highly expensive and time consuming. In the case of this study, the main objective is to look at the possibilities of implementing an on-board check-in and bag drop facility without the need for major change and cost. It has been determined that a driverless operation should not be considered in this case study.

The Tyne and Wear Metro currently operates with two metro cars coupled together. This is due to most platforms being unable to accommodate more than two cars (Nexus 2013). In order to keep costs down and the practicality of implementing such a system into the current Tyne and Wear Metro service, the design should attempt to avoid the necessity for a third metro car, remaining a two-car system. The aim is to minimise the amount of space required to facilitate a check-in service and prevent the capacity of the vehicle being profoundly affected by passengers not travelling to the airport.

There are no Tyne and Wear Metro vehicles solely dedicated to serve the Green Line. All vehicles must have the potential to operate for both Green and Yellow Lines. It is not possible to dedicate a set number of vehicles to Newcastle International Airport only. This information for our study removes the possibility of installing permanent facilities to the interior of the metro, affecting the capacity and full capability of the metro when the on-board check-in and bag drop facility is not in use. It will reduce capacity and the possibility of upsetting existing customers and deterring regular passengers from utilising the metro as a first choice to commute. Installing the check-in facility is intended to build upon the successes which already exist from the Tyne and Wear Metro and create a quality service suitable for passengers of all types. Consequently, a key feature for the designs has been to ensure that during the times in which the check-in desk is not in use, all equipment can be removed and stored away without obstructing large spaces and seating areas for passengers. The designs have been produced to allow the equipment to be removed conveniently and efficiently from the metro cars and transferred into other vehicles when required. Designs have been modelled concerning the equipment and layout currently utilised within the metro, limiting the degree and cost of upgrade required to all the metro vehicles and prevent equipment being required for every metro vehicle.

For each set of doors on board the metro, a manual pull down handle installed on one side. These handles are located next to the doors, at around 2000 mm up

Fig. 5 Manual emergency door handle on-board metro car 4067

from the floor of the vehicle. Within the centre of the metro, there are two of these mechanisms. It is an irrefutable health and safety requirement that the doors remain accessible at all times. New designs are developed considering this equipment. The location of this handle can be seen in Fig. 5.

Within location 2, at the centre of the metro, two containers extruded into the sides of the metro and contain fire extinguishers, as seen in Fig. 6. The two fire extinguishers should be accessible at all times during transit in the case of an emergency. The designs accommodate this health and safety factor to assure the correct number of fire extinguishers at all times.

Due to the centre of the metro sitting on a pivoting bogie, the slight rotation is inhibited at location 2, as seen in Fig. 7. The floor of the metro rotates relative to the rotation of the bogie. It is essential that equipment does not overlap between the rotating pivot and the stationary section of the metro as this can result in movement within the equipment making it unstable, unsafe and impractical.

For designs in which equipment is transported on and off the metro car, considerations have to be made to the dimensional limitations. The maximum height of equipment is 2016 mm (the height of the doors), while all equipment must be able to pass around the centre vertical handrail, providing 960 mm of floor space. Equipment that exceeds these dimensions will not be able to be transferred onto the metro unless dissembled first, which is impractical.

Nexus confirmed the possibility of installing 240 V plugs to the centre location of the metro vehicles. Currently, the metro utilises a considerable percentage of electricity due to the existing operation of the metro system. We considered designs based on the selection of equipment requiring a minimal power supply.

Fig. 6 Fire extinguisher casing on-board metro

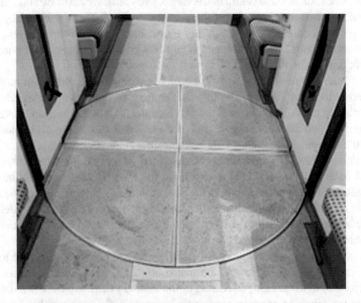

Fig. 7 Centre section of metro facilitating a pivoting bogie

Fig. 8 Areas where installation of check-in desk was considered

Currently, no storage exists on-board the Tyne and Wear Metro vehicles. The suggestion of utilising the unused driver's cabin for storage of equipment while in transit was dismissed due to safety issues.

6.7 Waiting Area

Through observing the centre area of the metro vehicle, the best location to position the check-in desk was in the centre archway of the vehicle, as seen in Fig. 8 location 2b. Positioning the check-in desk at the front of the centre compartment, as seen in Fig. 8 location 2a, would provide a greater area behind the check-in desk allowing more room to store luggage. However, there would be passenger queuing and door crowding issues when using the check-in desk.

Using location 2a would require passengers, waiting to check-in luggage, to stand with large baggage, potentially impeding clear access to board and exit the metro via the operating doors. Where there is a sudden surge of passengers intending to utilise the check-in and bag drop facility, a backlog of passengers can become a major issue as they are forced to move down the narrow aisles in order to fit onto the metro., This problem fails to address the passengers navigating luggage between closely packed seats and the potential for a slow boarding process. If passengers find difficulty in finding sufficient room, then boarding becomes a lengthy process and reduces efficiency of the service. The metro cannot leave the platform while a queue of people is extruding from the doors. In the case of passengers travelling to the airport, being unable to board the metro will not be accepted due to the importance of punctuality for flights. Incorporating the system in which specific metro vehicles serve certain flights, will make it more critical for passengers to board the required metro or face missing the opportunity to use the check-in facility and risk missing flights. It is essential to not only consider sufficient room for luggage storage as well as reserving space for passengers utilising the check-in facility.

The solution to these issues was addressed by positioning the check-in desk at the centre of the metro, allowing for priority seating and standing space for airport

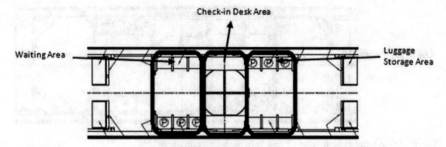

Fig. 9 Layout of check-in and bag drop facility

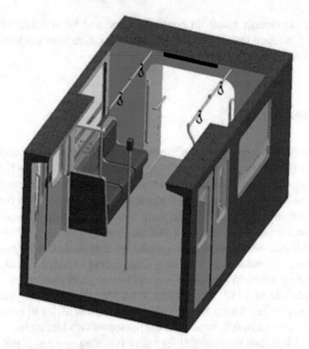

Fig. 10 Autodesk inventor display of design for airport waiting area

passengers, while retaining sufficient room for luggage to be stored. The basic layout incorporated into all the designs was designed into three sections, as seen in Fig. 9.

This arrangement allows passengers using the check-in facility to board the metro and move into the designated airport priority seating area. Passengers with luggage can be accommodated within this area, preventing obstruction to other passengers and reducing the potential of overcrowding around the doors. A representation model was developed as seen in Fig. 10, to demonstrate the design for this seating area.

By using this area of the metro for the check-in facility, there is minimal change. Passengers waiting to check-in will remain with their luggage which necessitates the need for sufficient floor space. Longitudinal seats not only open up space inside the

Fig. 11 Differing width of aisles between the centre of metro and standard seating

vehicle but are an ideal design for passengers being able to position luggage in front while seated.

The aisle along other seated areas on-board the metro vehicle provide a narrow width of just 600 mm while access to the longitudinal seating area provides a wider opening with an aisle width of 1200 mm, as seen in Fig. 11. This wider access to the seating provides ideal conditions for passengers moving through the metro with baggage.

Additional features added to the waiting area are basic and therefore cheap. Stickers will be added to the glass shields above the side boards which state 'Airport Check-In Area', as seen in Fig. 12, presenting the direction and efficiency for airport passengers upon boarding the metro. This feature would have to be applied to all metro cars as it is not a feature which could be removed. It is a low-cost addition and would not disrupt the current operation and layout of the metro when the check-in facility is not in use.

On the window above the longitudinal seats, located in the waiting area, there would be a sticker labelled as 'Airport Passenger Priority Seating', as seen in Fig. 13, providing additional assurance for passengers boarding the metro with luggage that there are available seats. This feature would also be applied to all metro cars and does not affect the operation of the metro when the check-in desk is not in use. The priority seating can remain for passengers travelling back from the airport and passengers using a metro vehicle travelling to the airport but without the check-in facilities on-board. Due to this only being priority seating, where there are journeys in which there is a low number of airport passengers, other passengers may use this seating area, assuring there is only a small section of the metro that is temporarily occupied.

Fig. 12 Airport check-in
area labels above side boards

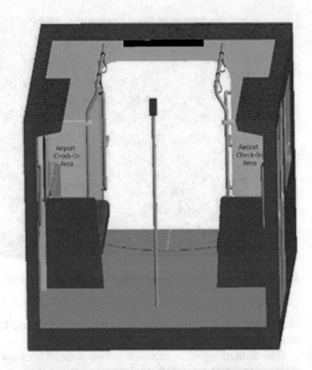

Fig. 13 Airport passenger
priority seating label

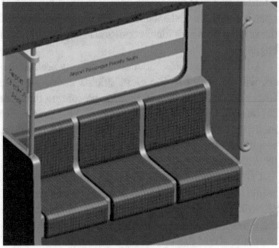

Fig. 14 Airport advertising within the waiting area

Currently, advertising is presented along the top corners of the metro. An effective way to promote this space would be to offer advertising instalments for Newcastle International Airport within the waiting area of the check-in facility, assuring the airport that advertising is being directed to relevant customers. It is a genuine incentive for the airport to help in funding and operating this service. A visual of how this advertising could be presented has been displayed in Fig. 14.

To further organise passengers within the waiting area, consideration must go towards utilising a ticketing system in which passengers collect a numbered ticket upon boarding. Passengers could be seated and then wait for the relevant number to be called before moving to the checking desk to hand over luggage preventing unnecessary queuing and crowding around the check-in desk while streamlining passenger dispersal along the metro vehicle.

6.8 Designs

Four designs have been considered (1, 2.1, 2.2 and 3) for the check-in desk and storage area of the check-in and bag drop facility and developed as a product design. The next section discusses these designs in detail.

6.8.1 Design 1

Design 1 has small permanent changes made to the centre of the metro car. These features can be packed away and removed and do not affect the normal operation of the metro if the check-in facility is not being utilised. When this design is in

Fig. 15 Full display of
Design 1 when in operation

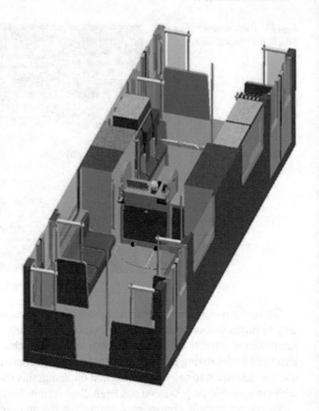

operation, the location 2b and the rear set of seats in the centre of the metro are
occupied.

A full depiction of the design when in operation can be seen in Fig. 15, while
the design when the check-in and bag drop facility is packed away can be seen in
Fig. 16.

Features of Check-in Desk

The check-in desk design is mobile so it can be easily removed and transferred to
different metro vehicles when required. The desk has been made mobile by adding
a set of wheels to the back of the desk. Handles are extruded from the check-in desk
at either side as to assist in manoeuvring the desk conveniently. An overall design is
displayed in Fig. 17.

In order for the check-in desk to be supported and stable on-board the metro,
a groove, mirroring the radius of the horizontal hand rail is positioned in location
2b. It is cut into the side of the check-in desk while a drop-down support has been
designed into the side of the desk. A knob is rotated to drop the support which causes
the surface of the desk to drop down. Carved into this surface is a groove with the

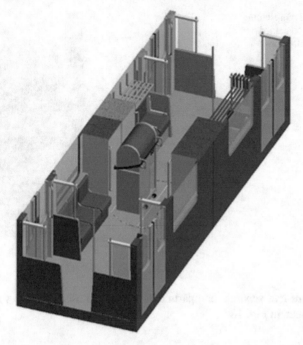

Fig. 16 Full display of Design 1 when packed away

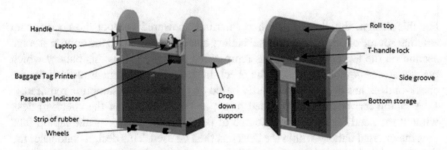

Fig. 17 Design 1 check-in desk 3D model

radius of the horizontal hand rail in which the support slots. These features can be seen in Fig. 17.

This design contains all the components required to operate a check-in facility within its storage and has two sets of storage. The top section is contained by a roll top function. In the front of this is a T-handle lock which is used to provide a locking system for the check-in desk. This design adds convenience by lifting back the roll top and pushing it down to the back of the desk. Beneath the roll top, there are several airport equipment products. The bottom section is contained by a set of doors with a set of cut out handles and a cam lock to secure the doors. Within the bottom storage

Fig. 18 Model representing
the equipment contained
within the check-in desk

area, there are five separate compartments which are used to store various larger
equipment, seen in Fig. 18.

Equipment

For this design, the check-in desk is internally powered without the need for an
external source of power. An external battery can be installed into the bottom storage
section on the top compartment, as represented in Fig. 18. It is this battery which
powers all the equipment within the check-in desk. The equipment utilised for the
check-in desk has been selected partly based on the power consumption required.

A laptop has been recommended to act as the computer of the check-in desk
without the need for an external source of power. If the battery of the laptop falls
low, the internal battery within the desk can then be used. This design condenses the
time in which the battery is powering the computer. The laptop would be fitted onto
a permanent mount fixed to the check-in desk workstation within the top storage of
the desk, displayed in both Figs. 17 and 18.

A barcode scanner scans the boarding passes of the passengers. For this design, we
selected the Xenon 1900 Barcode Scanner for passenger services. The Xenon 1900
is a tool used to gather data of tickets and boasts the ability to be applied to bag drop
check-in counters and boarding passes. This specific model was selected due to the
compact design, low operating power requirements and basic installation connecting
directly to a PC workstation via USB (Desko 2016). In conjunction with this device,
the scanner will be mounted on a stand which will be fixed to the workstation to
prevent movement during transportation of the desk. This device would be stored
in the top of the desk as seen in Fig. 18. The recommended mount to use with this

Fig. 19 Indicator system in
the check-in desk

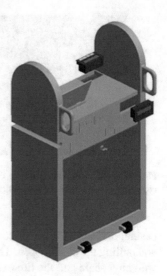

product is the Honeywell STND-22F00-001-4. The Xenon 1900 scanner would be
continuously powered by the check-in desk internal battery.

Stored inside of the bottom check-in desk is a weighing scale. In the case of this
design, a product known as the H305 bench scale has been used, as it is designed for
weighing airport luggage. It is a compact shape and can be stored inside the bottom
of the desk of the metro vehicle.

There must be an indicator display system for both the passenger and staff member.
The H305 Bench Scale used to weigh luggage for this design, can interface to all
Avery Weigh-Tronix indicator systems (Avery Weigh-Tronix 2015). We selected the
ZM205 model as it comes with both an operator display and passenger display unit
and can be easily connected to the weighing scale from the check-in desk. Both
products are compact with face displays of 200.7 × 108.7 mm. Rectangular cuts
are incorporated into the work station and in front of the check-in desk, as seen in
Fig. 19. They are bolted to the desk while the passenger display is covered by a clear
plastic screen to prevent damage, particularly when the desk is on-board the metro
but not in use. Both products have a low power consumption of just 150 mA at 12
VDC and are to be powered by the check-in desk internal battery.

For this design, the TK180 Baggage Tag Printer was selected due to its compact
design and 200 mm/sec high speed printing. A USB interface allows flexible integration, suitable for this application (Sita.Aero 2016). This product would be fitted
to a fixed mount on the work station of the check-in desk within the top storage of
the desk. This product has a low 24VDC power requirement and would be powered
by the check-in desk internal battery.

Due to the positioning of the check-in desk for this design, luggage would be
weighed and transferred to the fire extinguisher container at the centre of location
2b. The constant movement of luggage at this point can cause damage over time to the
screen containing the fire extinguisher. A loose strap hangs from the fire extinguisher

Fig. 20 Initial installation of the check-in desk with a fire extinguisher encased within its own compartment inside the check-in desk

casing, and this is required to be covered to prevent it from catching with luggage and pulling the casing open. Therefore, a shield was designed which fits between the weighing scale and the fire extinguisher casing and stored inside the check-in desk when not in use. To be able to accommodate this, the shield has been hinged down the middle so that it can be folded in half and packed away.

Due to the position of the check-in desk when installed, both fire extinguisher emergency containers are obstructed so two fire extinguishers are stored within the bottom storage of the check-in desk. Having the required amount of fire extinguishers while the metro is in transit complies with health and safety regulations.

To account for the fact that the check-in desk would obstruct access to one of the fire extinguisher cases in the centre of the metro when the desk is stored away, an emergency plastic cover was added to the back of the check-in desk to be used to access one of the fire extinguishers stored within the check-in desk. This fire extinguisher is encased within its own compartment inside the check-in desk, as shown in Fig. 20, and prevents the possibility of this feature being tampered with or stolen, allowing the correct number of fire extinguishers to remain accessible at all times.

Assurance is provided that the fire extinguisher casing on the back of the check-in desk is not accidentally damaged when passengers are standing in front of it as when the check-in desk is open, the roll top falls down in front of the casing thus protecting it.

The drop-down support is lowered onto the opposite horizontal hand rail on-board the metro to add further structure to the check-in desk, allowing for a secure fit and extra support, as seen in Fig. 21. Beneath this support, the weighing scale is fitted, and luggage passed through.

The shield is slotted into the drop-down support with two narrow slits machined at the end. The shield has two teeth which can mate with the slits in the support. By lowering the shield down through the horizontal hand rail, the teeth can slot into the slits of the support while the shield hangs over the fire extinguisher container while just coming into contact with the metro floor. This process can be seen in Fig. 22.

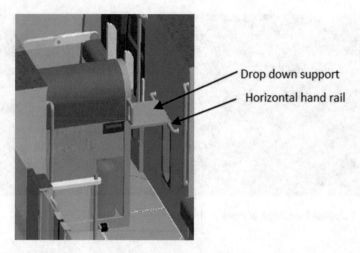

Drop down support

Horizontal hand rail

Fig. 21 Check-in desk supported horizontal hand rail

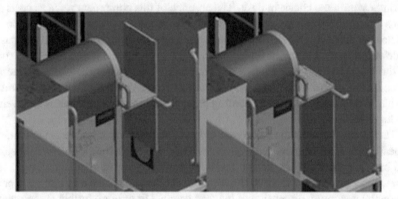

Fig. 22 Fire extinguisher shield setup

The weighing scale is removed from the check-in desk and placed on the floor of the metro in-between the side of the desk and the shield covering the fire extinguisher container. The design of the desk is such that when the weighing scale is fitted, the width of the scale, shield and desk accommodates the full width of the metro, creating a compact fit which stabilises the check-in desk and tightly fits the scale in place. The visual setup of the check-in desk once successfully installed can be seen in Fig. 23. Fitted to the lower side of the check-in desk and shield is a strip of rubber which sits parallel to the H305 weighing scale. This adds further assurance by increasing the friction between the components and reducing the chance of movement while in transit.

When the check-in desk is opened, the roll top storage can be unlocked, lifted up and pushed down the back of the desk. This exposes a work surface, with all the equipment required to operate an airport check-in facility.

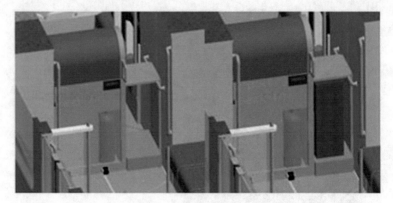

Fig. 23 Complete installation of check-in desk on

Passengers check in online, before boarding the metro and removing the need for a printer within the check-in desk. The absence of this equipment reduces costs and saves space on an already highly condensed work surface. Checking in online is a service provided by most airlines where passengers can print their boarding passes or download onto smartphones. Incorporating this requirement also intends to speed up the baggage drop process on-board the metro and reduce the waiting time and potential for extensive queuing. Upon boarding the metro, the staff member operating the bag drop facility scans the boarding pass, checks the passport details and weighs the luggage.

The section between the horizontal support and the weighing scale is facilitated as a measurement for the maximum allowable size of luggage for the on-board check-in and bag drop service. The design has been produced to accommodate 'large' suitcases, as represented in Table 2. However this feature could be easily altered depending on the maximum dimension of luggage deemed fit through further assessment of the check-in facility. In the case of this study, luggage greater than a 'large' suitcase would occupy too much space if a significant amount of luggage is d to be stored on-board the metro.

Upon passengers checking in baggage, the luggage is moved to the storage area while the passenger is free to be seated on the metro.

The first step in closing the check-in desk is to switch off all the related equipment on the desk followed by the internal battery. The shield protecting the fire extinguisher casing would be lifted up and removed from between the metro wall and weighing scale. The equipment would be stored inside the check-in desk. To be able to accommodate this, the shield has been hinged down the middle so that it can be folded in half and packed away. The H305 Bench Scale would be detached and stored inside the desk. The lock connecting the check-in desk to the horizontal hand rail via the handle of the desk is then to be unlocked by the member of staff and stored in the desk. Following unlocking, the doors of the storage unit within the check-in desk would be locked shut. The fold down support, which sits on the horizontal hand

Fig. 24 Check-in desk locked away

rail on the side of the metro, would be lifted and locked in place against the side of the desk. The roll top of the desk would then be pulled over the top of the check-in desk and locked thus exposing the fire extinguishers emergency casing on the rear of the desk, containing all equipment safely inside of the desk, as seen in Fig. 24.

The check-in desk can be prised away from the horizontal hand rail, situated within the side groove of the desk, and rotated so that the front of the check-in desk is facing the centre compartment of the check-in and bag drop area at location 2b. A strap, which can be stored inside the check-in desk, is then hooked onto the end of the horizontal hand rail, threaded between the handles of the desk and connected to the opposite end of the hand rail. This secures the desk in place and prevents it from tipping over in the event of emergency braking or an impact, as seen in Fig. 25. With the check-in desk stored away in this way, this allows for plenty of space (965 mm) for passengers to move around the desk. Following this, all equipment for the check-in and bag drop facility has then been successfully packed away which does not remove any seating space on the metro. The interior can operate in the same way where only a very small section of the metro is accommodated by the facility. Figure 25 shows the final configuration of the check-in desk when stored away.

The Storage Area

To accommodate a greater capacity for luggage in the storage area, two sets of baggage racks have been designed above the longitudinal seats which sit opposite each other. These baggage racks would be a permanent addition to all metro vehicles.

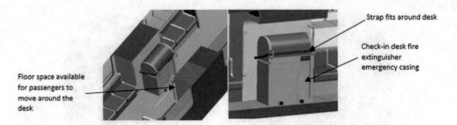

Fig. 25 Check-in desk when not in use

Fig. 26 Baggage racks

The Tyne and Wear Metro currently lacks any storage space for baggage. Installing this feature would help improve an all-round customer satisfaction, not just specific to airport passengers. The dimensions of the baggage racks have been designed to fit around the current layout of the metro while assuring there is sufficient room for passengers to s utilise the seats. The addition of luggage racks would not affect the capacity of the metro vehicle, improving the facilities available for all metro passengers.

The designs for baggage racks differ from each other to accommodate the manual emergency door handle, located above one side of the storage area. The luggage rack above the longitudinal seating does not facilitate the emergency door handle and is designed to provide two levels of storage across the full length of the seating, as seen in Fig. 26.

The baggage rack is connected to the side of the centre compartment of the check-in area via two levels of six extruding rods. The opposite end is supported on the horizontal structure supporting the vertical hand rail above the side board of the storage area. This provides a strong structure to support heavy luggage. The rack installed onto the metro can be seen in Fig. 27.

The design of the baggage rack located above the longitudinal seating facilitates the emergency door handle has been produced to retain clear access to the handle, seen in Fig. 28. This reduces the capacity of luggage in the baggage rack but is

Fig. 27 Racks installed onto
the metro car

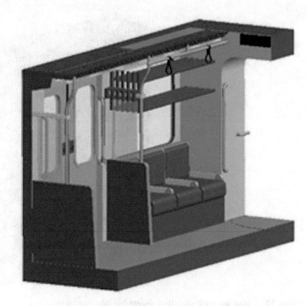

Fig. 28 Design of luggage
rack above longitudinal seats
with the emergency door
handle

retained to provide an increase in overall baggage capacity adhering to health and
safety requirements.

This baggage rack cannot be supported on the horizontal structure supporting the
vertical hand rail as the emergency door handle needs to be clear. An alternative
design has been considered. Like the first baggage rack design, this rack will be
connected to the side of the centre compartment of the check-in area via two levels
of six extruding rods. The rack will be supported by vertical rods extruding from the
back and connected to the ceiling of the metro. The dimensions of this rack enables
the rods to fit behind the light casing and venting system for connection to the ceiling

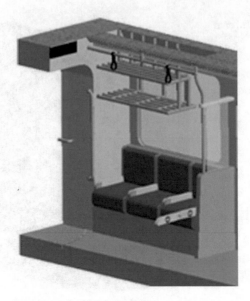

Fig. 29 Luggage rack installed above longitudinal seats

and access to the lights for servicing. The rack installed onto the metro can be seen in Fig. 29.

Due to the location of the check-in facility, passengers could be able to access the opposite end of the check-in area by boarding side B of the metro. It is important that this area is not accessed by passengers to assure the security of the luggage. To accommodate this requirement, it is proposed that all metro cars would be fitted with 'fold out arms' installed onto the longitudinal seats in the storage area, see Fig. 30. These fold-out arms are installed on the spacing between the seats so not to affect the capacity of the metro when the check-in and bag drop facility is not in use. To implement the barrier and to prevent access to the storage area, a doubled jointed arm rest is located at the end of the longitudinal seating. It is extended outwards, lying flat and parallel to the base of the seat, as seen in Fig. 31. When both double joined arms are folded out, a complete barrier falls across the width of the storage area, indicating that access is denied.

No entry signs add clarity for passengers that the storage area is a restricted area and not to be accessed.

The longitudinal seating within the storage area of the check-in and bag drop facility is to store checked in luggage. Supports have been installed to prevent the stack of luggage from collapsing. Single jointed 'fold up arms' lift backwards but constrained to fall back no further than 90° from the face of the seats without a locking mechanism.

Stored within the 'fold up arms' are straps which can be pulled from within the arms and clipped to the luggage racks to provide further support.

Fig. 30 Set up of double jointed fold out arm to act as barrier

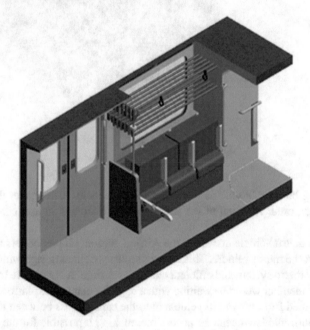

Fig. 31 Fold out arms lifted up

On opening the check-in and bag drop facility, the storage area must be prepared for use. This design provides a simple setup as the double jointed 'fold out arms' can be dropped down to create a barrier to prevent entry by passengers into the storage area.

Following luggage being checked in, the 'fold out arms' are lifted up and the luggage is placed behind them, as seen in Fig. 32. Baggage should be stacked on the seats until the luggage racks prevent room for further luggage to be stored. Once the stack of luggage exceeds the height of the folded out arms, the straps can be pulled from within the arms and clipped to the luggage racks.

Fig. 32 Straps from arms
clipped to the luggage rack

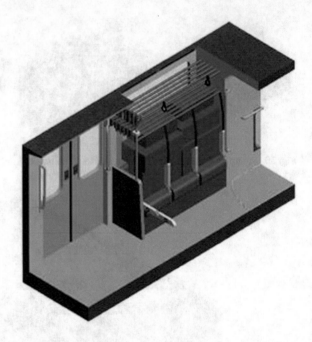

Following both seats reaching the maximum capacity of luggage, the baggage racks are then used. A visual of the storage area, set up and in use, can be seen in Fig. 33.

Once the metro vehicle arrives at the Airport station, all passengers are allowed to disembark the metro vehicle. This assures security of the luggage whilst allowing for an empty metro system and efficient change over time. A Newcastle International Airport staff member would be waiting with a baggage cart on the platform. Luggage is then unloaded from the metro vehicle onto the cart. Access between the platform and airport terminal is wheelchair accessible, making it possible for the loaded cart to be transported to the Airport building without significant effort.

Once the equipment is stored away, passengers travelling from the airport can board the metro and utilise the luggage racks. This provides assistance for a baggage free journey. Figure 34 shows a display of the storage area after luggage is offloaded, and all features are packed away.

Based on the shape and size of the storage areas for the luggage, an estimation could be made of the baggage capacity of Design 1. Set dimensions for each type of luggage that may be used by passengers vary in terms of the height, size, width and volume. Estimation for the capacity of this design is displayed through a calculation of the volume of each section of the storage area divided by the volume of the varying luggage types used for this study. The value of luggage was in each case rounded down to zero decimal places. This is displayed in Table 3.

Considering the worst case scenario in which all passengers' board with large luggage, this would give the facility a baggage capacity of 18 bags per journey.

Fig. 33 Storage area setup
and with a full capacity of
luggage

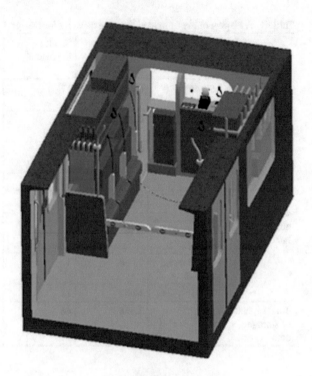

Fig. 34 Storage area packed
away

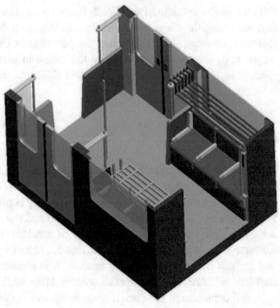

Table 3 A display of the estimated baggage capacity for Design 1

		Volume capacity (m^3)	Small luggage	Medium luggage	Large luggage
Storage side without emergency handle	Seating storage	0.835	17	11	7
	Lower rack storage	0.249	5	3	2
	Upper rack storage	0.186	3	2	1
	Total	**1.27**	**25**	**16**	**10**
Storage side with emergency handle	Seating storage	0.835	17	11	7
	Lower rack storage	0.165	3	2	1
	Upper rack storage	0.088	1	1	0
	Total	**1.088**	**21**	**14**	**8**
Total capacity of storage area		**2.358**	**46**	**30**	**18**

With the metro operating from South Hylton to Airport for approximately 16.5 h per day, assuming the on-board check-in and bag drop facility operates on two metro journeys per hour, this system has the potential to facilitate the movement of 594 'large' suitcases per day. Assuming the check-in service operates on every metro journey to the airport, this design could facilitate the movement of 1485 'large' suitcases per day.

6.8.2 Design 2.1

Design 2.1 has been created in which no permanent changes need to be implemented into the current design of the centre of the metro. This allows for all equipment to be removed when the check-in and bag drop facility is not utilised on-board the metro. When this design is in operation, location 2b and the storage area is occupied. This design aims to offer an alternative way of handling the luggage whilst looking at alternative equipment and operations that could be implemented. It should be noted that for this design to efficiently operate, it is recommended that two members of staff are on-board, with one operating behind the check-in desk and the other in front.

A full depiction of the design when in operation can be seen in Fig. 35, while the design when the check-in and bag drop facility is not in use can be seen in Fig. 36.

Fig. 35 Design 2.1 when in
operation

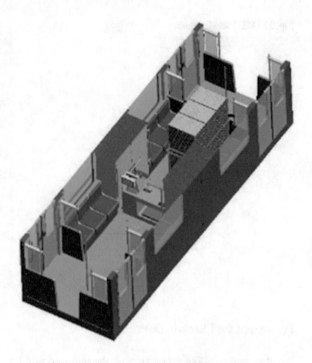

Fig. 36 Design 2.1 when
packed away

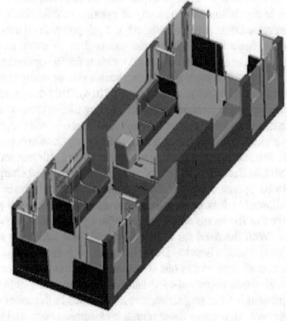

Fig. 37 MD1 service desk

Features of the Check-in Desk

An alternative way of incorporating the required facilities to operate a check-in desk is to purchase a bulk quantity of existing mobile check-in kiosks. The MD1 Mobile Service Desk, as seen in Fig. 37, is equipped with all the capabilities of a full check-in desk, including boarding pass and bag tag printers, an adjustable passenger facing LCD screen and keyboard and monitor for the operator.

In addition to this, weighing scales can be easily connected to the MD1 through its external input panel (Embross 2016). This design considers the opportunity for passengers to be able to obtain boarding passes thus expanding the customer base for passengers who can use the on-board check-in and bag drop facility. By utilising this service desk, it reduces the amount of separate components required on-board the metro and assists in a more user friendly and efficient service. Furthermore, the MD1 Mobile Service Desk contains its own battery and charging facility thus allowing it to be operated without the need for an external power source for up to 10 h. This allows for desk to be able to operate in the event of a power malfunction from the plug of the metro which in turn offers greater reliability of the facility.

With the need for a weighing scale and indicators to be connected through the MD1 Mobile Service Desk, it would be a disarranged setup if the equipment was not contained within one structure. This would also create a health and safety hazard with wires being exposed and difficulty in being able to lock down equipment to prevent it from projecting down the aisles in the event of emergency braking or an impact. To remove these issues, a casing has been designed in which the MD1 desk can be fitted into whilst the weighing scale and indicators are also incorporated into the design. This would allow for more efficient handling of the equipment as all

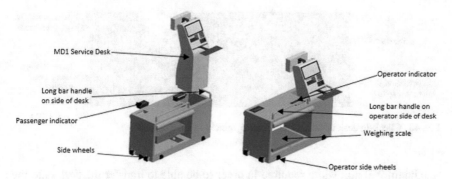

Fig. 38 Explosive view of Design 2.1 check-in desk

Fig. 39 Check-in desk casing and shutter mechanism

the required components are contained within the desk. The MD1 desk fits into the casing as seen in Fig. 38 whilst the bare casing can be seen in Fig. 39. The service desk is padlocked internally to keep it fixed to the casing and prevent theft.

Within the container for the service desk, a large opening accommodates the H301 weighing scale. A rectangular intrusion to the base of the opening is implemented which allows the scale to sit in, reducing the portion of the scale exposed. From here, the wiring can travel from the scale and through a small hole that passes through into the section containing the MD1 desk.

The indicators used with the H305 Bench Scale are the Avery Weigh-Tronix indicator systems. Similar to design 1, both passenger and operator components are fitted into rectangular cuts which have been designed into the work station of the casing for the MD1 desk. These are bolted to the desk whilst both passenger and operator display are covered by a clear plastic screen to prevent damage to the indicators.

Two metal roll down shutters cover the weighing area of the check-in desk as to protect the scales and allow for a storage area to store equipment in when the desk is not in use. This is functioned by installing two horizontal axles within the weighing area of the desk. The metal shutters fit around these axles and are able to roll up and down. The base of the shutters is fitted with a small T-handle lock which can allow the shutters to lock into the inner base of the check-in desk.

The casing for this design is mobile to allow it to be transferred to different metro vehicles when required. The casing creates a check-in desk which is wider than the

Fig. 40 Check-in desk installed on the metro vehicle

maximum 960 mm width required in order to be able to transfer the desk onto the metro. To accommodate for this, wheels have been added to the side of the desk thus allowing it to be lifted up and pulled onto the metro vehicle. A long bar handle has been added to the opposite side of the check-in desk to assist in manoeuvring it conveniently. Further to this, wheels and a long bar handle have been added to the operator side of the desk. This allows for improved manoeuvrability when on-board the metro in positioning the check-in desk.

Equipment

Using the wheels and handle on the side of the casing, the check-in desk can be transferred onto the metro. Upon the check-in desk entering the centre of the metro, the desk is then rotated such that it is adjacent to the walls of the metro and the 'Airport Check-In Desk' text on the desk is facing towards the waiting area. From here, the desk is to then be pulled back into place within location 2b, using the wheels and handle on the operator side of the desk.

The check-in desk is able to fit in the centre compartment of the check-in and bag drop area whilst remaining in front of the fire extinguisher containers on-board the metro. This allows for clear access to the fire extinguishers at all times without the need to store extra fire extinguishers within the check-in desk which saves money and weight. To allow for sufficient space for the members of staff to operate behind the check-in desk, the desk is stored at the front end of the location 2b whilst avoiding overlapping the pivoting floor. Upon the positioning of the desk being complete, two sets of locks have been fitted around the long bar handles and horizontal hand rails seen within the vehicle. This prevents the desk from projecting down the aisle of the metro in the event of emergency braking or an impact. This initial installation can be seen in Fig. 40.

Covering the monitors or the check-in desk is a tarpaulin elastic cover that fits over this equipment. Upon the opening of the check-in desk, the cover is removed and placed on the seats within the storage area. The passenger monitor and operator keyboard of the MD1 Mobile Service Desk and then extended outwards ready for use.

The two metal roll down shutters cover the weighing area of the check-in desk are to be unlocked and rolled up thus exposing the H305 Bench Scale. Following the

Fig. 41 Check-in desk
successfully setup

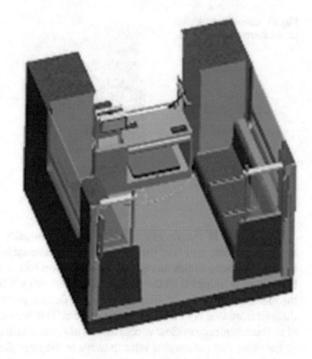

processes mentioned, the check-in desk would be successfully setup to look as seen in Fig. 41.

Upon boarding the metro, passengers would move through the waiting area and towards the front of the check-in desk. From here, boarding passes can be obtained and luggage can be weighed. The opening containing the weighing scale could act as a determinant as for whether luggage is of a suitable size to be checked-in using this facility. Alternatively, the smaller height of the work station of the check-in desk can allow for the potential of larger luggage to be passed over it. For this design, however, it is assumed that baggage greater in volume than 'large' luggage is not acceptable for use. Upon luggage being approved, the baggage label and be printed and applied. From here, the luggage can be pulled through into the storage area whilst passengers are free to be seated on the metro.

When closing down the check-in desk, the desk would be unplugged from the metro power source followed by the mobile service desk being logged off and shut down. Other equipment from the check-in and bag drop facility is placed on the weighing scale within the check-in desk. Following this, the metal shutters are pulled down closed and locked shut. When this is done, the shutter displays the message 'closed' which provides a clear sign to passengers that the desk is not currently in use. The adjustable keyboard and passenger monitor of the MD1 Mobile Service Desk is then folded down, and the tarpaulin elastic cover is placed over the top of the service desk thus protecting it from being damaged. Once packed away, the check-in desk will look as seen in Fig. 42.

Fig. 42 Check-in desk when
closed down

The desk is to remain in the position as it is initially installed. This allows the fire extinguishers on-board the metro to remain accessible to passengers. No seating is accommodated in this storage of the desk. However, a small section of the metro floor space is occupied by the desk (around 0.65 m^2) whilst the internal operation of the metro must be altered slightly. With this design, passengers are not able to pass through from side A to B of the metro vehicle. This would not be perceived as much of an inconvenience as clear access to a set of doors is still available to all passengers. In the event that passengers want to move to the opposite end of the metro, access can be simply achieved though departing the metro and walking to the next set of doors on the opposite side of the metro.

The Storage Area

To prevent passengers accessing the storage area via the opposite side from the waiting area, an expandable safety barrier has been designed. The barrier can be collapsed to be stored within the weighing section of the check-in desk when not in use whilst being able to expand an utilised as a barrier. The end support of the barrier can slot onto the end of the side board at the edge of the storage area, strapped to the vertical hand rail and then expanded outwards to be connected to the end of the opposite side board. This design was inspired by the Titan Expanding Barrier (Workplace Products 2016), as seen in Fig. 43. The design produced t can be seen in Fig. 44.

For this design, a roll container is used to store the luggage. This utilises the vast floor space within the centre of the metro, due to the use of the longitudinal seating, avoiding the need for any adaptations to the current components on-board the metro. Luggage is able to be safely and securely stacked up within the container and hence made for more efficient handling of the luggage.

The initial dimensions of the roll container were designed to be able to transfer and manoeuvre the container on and off the metro vehicle. To accommodate this, the depth was assured to be no greater than 960 mm whilst the height of the container was

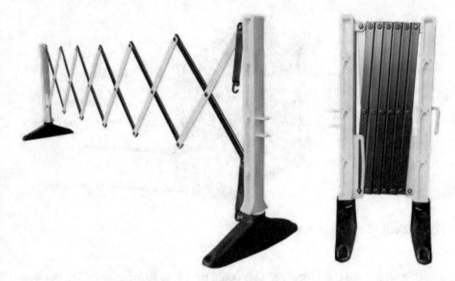

Fig. 43 Titan expanding safety barrier (Workplace Products 2016)

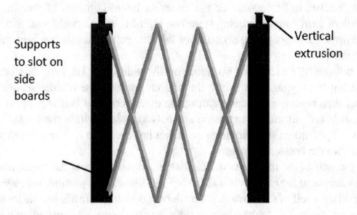

Supports
to slot on
side
boards

Vertical
extrusion

Fig. 44 Model produced of expanding safety barrier

based less than the height of the doors of the metro vehicle (2016 mm). The length of the container was then judged by the length of the length of the longitudinal seats within the storage area (1710 mm), such that it was of a similar length to this, as to assure the roll container did not interfere with the staff member operating the check-in desk and to avoid the wheels overlapping onto the pivoting floor. This design created the maximum storage capacity of the check-in desk whilst retaining flexible and efficient operation on the Tyne and Wear Metro.

With the roll container being under two metres tall, access to the storage is via a door on the side. With this considered, it would not be safe to stack several loads of luggage on top of each other at two meters in height as the baggage can become

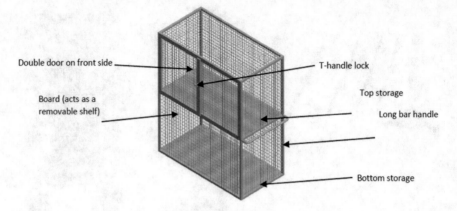

Double door on front side

T-handle lock

Board (acts as a
removable shelf)

Top storage

Long bar handle

Bottom storage

Fig. 45 Design 2.1 roll container

unstable within the roll container. This could then lead to luggage falling out from the container upon opening the door. Furthermore, customers may be dissatisfied as luggage checked in first would be put under an intense amount of weight, from a high stack of heavy luggage being stored on top of it, which could lead to damaged goods within the baggage. To account for this, the roll container has been split into two sections.

The bottom half facilitates no doors on the side of it. Luggage is stored here by lowering the luggage down into the bottom section. The middle section of the container then has a flat extrusion around the inner perimeter half way up, as seen in Fig. 46. A board sits on this extrusion and acts as a shelf within the container. This board is hinged down the middle as so it can be folded up to allow luggage to be dropped into the bottom storage.

The top half of the roll container facilitates a double door at the front side which provides access to luggage stored at the top section. At this section, luggage can be stacked on the shelf of the container. The doors accommodate a T-handle lock which provides an ergonomic feature help in opening the door whilst providing a locking system to safely contain the luggage within the roll container. The roll container designed can be seen in Fig. 45.

This roll container has been designed so that it is mobile, easily transferred on and off the metro vehicle. To accommodate this, four heavy duty of swivel castors have been implemented to the base of the container. Integrated into the wheels of the roll container are brakes which can be applied when on-board the metro, as to prevent the container from moving while in transit. A long bar handle has been fitted at the end of the container to provide a suitable grip while manoeuvring it, as seen in Fig. 46.

The walls of the container are caged, as seen in Fig. 46, to ensure it is light weight and easy to move, providing the staff on-board with the metro with a clear vision of

Fig. 46 Design 2.1 roll
container base

the luggage at all times. The use of minimal material for this design also reduces the cost of manufacturing the product.

At the first station to the airport, South Hylton, an empty roll container is transferred onto the metro vehicle. The container is positioned flat against the left-hand side of the longitudinal seating with the longer side of the container sitting parallel to the seating. The double doors of the container are facing away from the longitudinal seats to facilitate clear access to the storage by the member of staff operating the check-in service. Once the container is correctly positioned, as seen in Fig. 47, the brakes of the swivel wheels are applied thus locking the roll container in place.

A lock is then applied to the long bar handle of the roll container and the vertical hand rail positioned on the edge of the centre compartment of the check-in and bag drop facility. A strap is also fitted around the vertical hand rail positioned above the side board adjacent to the longitudinal seating and passed through the caging of the roll container, as seen in Fig. 47. These added features provide additional safety and stability to the container as to prevent it from projecting down the aisle in the event of emergency braking or an impact. It is particularly important for this to be considered for this design as when the container is full of luggage, it becomes a heavily condensed weight on a set of wheels. The brakes and locks, therefore, account for this.

From within the check-in desk, the expandable safety barrier is removed and installed. The end support is slotted down the side of the side board. Present on top of the support is a vertical extrusion in which an adjustable strap is fixed to it. This

Fig. 47 Roll container
locked up in position during
operation

Fig. 48 Expandable barrier
set up

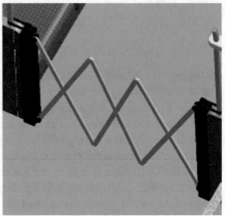

strap is tied around the vertical hand rail above the side board and tightened as to fix the support of the barrier. The opposite support is then pulled out and fitted into the end of the opposite side board. A strap is attached to this support as well which can be tightly fitted around the vertical hand rail. The straps prevent the barrier from pulling out from the side board whilst the ingrained slot of the supports which fit around the side boards provides a casing either side to prevent the barrier from falling forwards or backwards. This set up can be seen in Fig. 48.

Following these installation processes, the storage area setup is complete which is displayed as seen in Fig. 49.

Fig. 49 Design 2.1 storage
area in

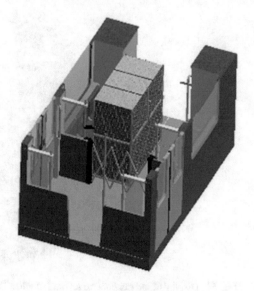

Once an item of baggage has been checked in, the luggage would be placed on the
seating opposite the roll containers. The staff member would then open the double
set of doors of the container and lift back the board enclosing the bottom section of
the storage. From here, the luggage can be lifted up and lowered into the roll container. Following the maximum capacity of the bottom storage being accommodated,
luggage would then be stacked on top of the board. Once the maximum capacity of
the roll container has been reached or the final station before the airport has been
passed, the doubled doors are locked shut to safely secure the luggage in place.

On arrival at the airport, all passengers depart the metro first. Following this,
the barrier is removed by disconnecting the straps from the vertical hand rails and
pulling the supports together thus collapsing the barrier. It can then be stored within
the check-in desk. The lock and strap are removed from the roll container, and the
brakes on the wheels are removed. A member of staff, employed by the airport, waits
for the roll container at the Airport Station. This staff member is required to board the
metro and manoeuvre the roll container off the vehicle and onto the platform. From
here, the roll container is wheeled away into the airport. The metro can then travel
away from the airport without a roll container on-board which therefore prevents the
container occupying valuable passenger space, and thus this area of the metro can
operate as usual.

With this design, a separate mode of transport would be required to transfer the
roll containers back to South Hylton. If a sufficient number of the roll containers are
bought in by Nexus, these containers can be accumulated at the airport and transferred
via a heavy goods vehicle back to South Hylton by the end of the day. This would
maximise capacity for passengers while the metro is operating.

In the case of this design, the safety barrier is simply collapsed down and stored
within the check-in desk. The roll container is then removed from the metro vehicle,

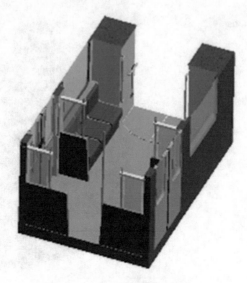

Fig. 50 Display of the same metro layout left when Design 2.1 storage away is packed away

Table 4 A display of the estimated baggage capacity for Design 1

	Volume capacity (m³)	Small luggage	Medium luggage	Large luggage
Bottom storage of roll container	1.035	21	13	9
Top storage of roll container	1.017	21	13	9
Total capacity of roll container	2.052	42	26	18

leaving the storage area in the same setup as the current metro design, as seen in Fig. 50.

As for Design 1, based on the shape and size of the roll container, estimation could be made as to the baggage capacity of Design 2.1. The dimensions used show set dimensions for each type of luggage however the height, size, width and volume of each individual luggage used by a passenger can vary. Therefore, an estimation for the capacity of this design is obtained through a calculation of the volume of each section of the storage area divided by the volume of the varying luggage types used for this study. The value of luggage was in each case rounded down to zero decimal places. This is displayed in Table 4.

Considering the worst case scenario in which all passengers' board with large luggage, this would give the facility a baggage capacity of 18 bags per journey. This provides a capacity similar to that of Design 1, however, this design requires fewer components and no permanent additions to the facilities on-board the metro. This, therefore, means this system has the potential to facilitate the movement of 594 'large' suitcases per day. Assuming the check-in service operates on every metro journey to the airport, this design could facilitate the movement of 1485 'large' suitcases per day.

6.8.3   Design 2.2

An alternative way of operating and storing the equipment for Design 2.1 was considered in which the roll container could be transferred back on the metro from the Airport Station. This would eliminate the need for alternative transportation of the container back to South Hylton thus saving money. This design would retain the same design for the check-in desk and foldable barrier. It would demonstrate a new design to the roll container in which the check-in desk can be stored within the roll container.

Features of the Roll Container

The general design of the roll container is kept the same. However, the dimensions of the container have been altered slightly. The container can fit into the centre compartment of the check-in and bag drop area. This centre compartment has a lower ceiling height of 1940 mm, and the height of the container would be reduced. The width and depth of the container remain the same. This reduces the luggage capacity that the roll container can store, nonetheless, the features of the board that sits within the container and the double door remain the same.

An added feature to this design is a side that can drop down to act as a ramp. Upon the check-in equipment being packed away, the ramp would be activated and the check-in desk pulled onto the roll container. Due to the height of the frame of the container being almost two metres in height, dropping down the whole side of the roll container would be impractical due to insufficient room. Furthermore, a ramp as long as this one would undergo intense forces when the check-in desk is being pulled along the ramp. To account for this, the side of the roll container is designed into two parts. The top section of the roll container contains a door, equipped with a T-handle lock. The bottom section has been reduced in height, hinged at the base and therefore able to drop down as to act as a ramp. The door at the side can be locked through the use of the T-handle lock. This door overhangs the bottom half of the roll container, and when the door is shut, the ramp remains in place. Upon opening the door, the ramp is then able to be dropped down. The functionality of this feature can be seen in Fig. 51. By reducing the height of the bottom section of the roll container and creating two openings to the side of the container, a ramp of just 600 mm in length can be utilised. This way, it is sturdy and more compact, while improving access to luggage.

The ramp of the container has been designed so that the tip is narrow. This allows the ramp to be as flush to the floor of the metro as possible when dropped down. The ramp has a similar caged design to the rest of the roll container. In the cage are two

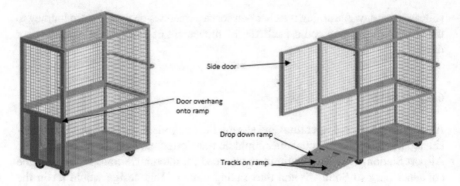

Fig. 51 Design 2.2 roll container with drop down ramp

tracks to create a smooth surface for the wheels of the check-in desk to follow. The design of the ramp can be seen in Fig. 51.

The extrusion around the inner perimeter of the roll container, in which the hinged board sits, does not exist at the side of the container, as seen in Fig. 51. This allows the board to operate in the same way but allows open access at the side of the container for the check-in desk.

The same mobility features are present for this design of the roll container; however, the long bare handle has been switched to the opposite side as to accommodate the ramp.

Due to the position in which the roll container is stored on-board the metro, new locations will need to be considered for the emergency fire extinguishers. It is suggested that these are located with one at the front end of the metro, at side A, and the other at the opposite end, at side B.

The Storage Area

Installing this design of the roll container involves exactly the same processes. The container is positioned at the same place, however, with the handle being on the opposite side of the container, the lock is fixed around the handle and attached to the vertical hand rail above the side board whilst the strap is fixed around the vertical hand rail at the edge of location 2b and passed through the caging of the container. This can be seen in Fig. 52.

The way in which the roll container is operated when the metro is in transit does not differ to Design 2.1. The side door and ramp should remain closed as to not disrupt the stack of luggage. The way in which luggage is stored onto the container is the same.

Following the arrival at Newcastle International Airport, the roll container is to be transferred off the metro and into the airport in the same way as described in Design 2.1. The difference is for the airport staff members to wait at the platform to transfer

Fig. 52 Installation of the
roll container for Design 2.2

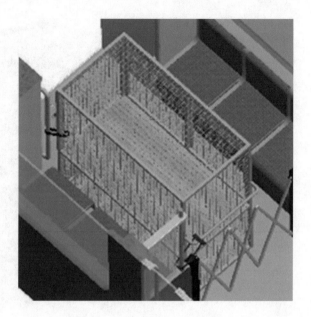

the luggage with an empty roll container. Upon the arrival of the metro, the roll container on-board, which contains the checked in luggage, can then be transferred into the airport while the empty roll container is transferred onto the metro. This empty roll container is then stored in the centre compartment, and all equipment from the check-in and bag drop facility is loaded onto it. The checked in luggage within the container, being unloaded and passed to airport security, the now empty roll container would be transferred back to the Airport Station to await loading onto the next metro service, accommodating the check-in and bag drop facility.

On arriving at the Airport Station, the barrier would be collapsed down followed by the roll container being transferred off the metro and the new empty container being transferred on. The new roll container would then be positioned in the centre of the storage area of the metro. Firstly, the hinged board separating the top and bottom storage of the container is folded back and lifted up and out from the extrusion. The side door of the container would be unlocked and opened up thus allowing for the ramp to drop down to the floor. The check-in desk is rotated 90°, such that the desk is aligned with the side entry of the container, and pulled onto the roll container via the drop down ramp. Following this, the barrier is then stored within the roll container, and the ramp is raised up and the door closed and locked shut. By avoiding the barrier being stored within the check-in desk, it removes the chance of the barrier damaging and scuffing the weighing scale. The roll container with all equipment stored inside can be seen in Fig. 53.

The roll container, containing the check-in equipment, would then be transferred into the centre compartment of the check-in and bag drop area of the metro at location 2b. It is assured that the wheels of the roll container do not overlap the pivoting floor. Once correctly positioned, the brakes on the wheels are applied to fix the container

Fig. 53 Roll container with
equipment stored within it

in place. Due to the size of the roll container, complete accommodation of location
2b, on the pivoting floor, is occupied by the container. To assure passengers do not
attempt to move between the metro and the roll container, two panels have been
designed which sit within the vertical hand rails positioned at the ends of location
2b. These panels also provide a barrier as to prevent the container from moving down
the metro in the event of emergency braking or an impact.

The barriers are designed around the shape and positioning of the roll container.
The first barrier is a flat rectangular shape. One end is curved such that it can be
clipped around vertical hand rail. At either end of the barrier, a cut is implemented
into the base as so the barrier can then be pushed down and slotted onto the bottom
section of the hand rail. The second barrier is of a unique shape. Due to the size of the
roll container, although the wheels are contained within the pivoting floor, a section
of container structure overhangs. Therefore, this barrier has been shaped such that it
curves outwards to tightly fit around the side of the roll container. The ends of the
barrier are then equipped in the same way as the first thus allowing it to be fitted
onto the vertical hand rails. Displayed on the front of these barriers is the message
'Keep Out' which provides clear instructions to the passengers that this section of
the metro is not available for use. These designs can be seen in Fig. 54 which also
displays the final setup of the check-in and bag drop equipment when stored away.

This design would not reduce the seating capacity of the metro. However, location
2b would be completely out of bounds for passengers to stand and move through.
With this section facilitating the pivoting tractor bogie, and therefore obtaining slight
rotation of the floor, and being of a smaller height without any windows, it would

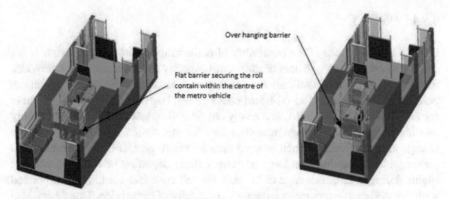

Fig. 54 The check-in and bag drop facility packed away for Design 2.2

Table 5 A display of the estimated baggage capacity for Design 2.2

	Volume capacity (m^3)	Small luggage	Medium luggage	Large luggage
Bottom storage of roll container	0.696	14	9	6
Top storage of roll container	1.017	21	13	9
Total capacity of roll container	1.713	35	22	15

be seen as the most unpopular place to stand, and therefore the unavailability of this area will not greatly affect customer satisfaction negatively.

As previously considered, based on the shape and size of the roll container, estimation could be made as to the baggage capacity of Design 2.2. The same method is used to measure the capacity of this design, as in Design 1 and 2.1, as to create an estimation for the capacity of this design. The results are shown in Table 5.

Considering the worst case scenario in which all passengers' board with large luggage, this would give the facility a baggage capacity of 15 bags per journey. This design provides a capacity smaller than Design 1 and 2.1. However, this design avoids the need to operate a separate mode of transport to transfer the roll containers back to South Hylton Station. This system has the potential to facilitate the movement of 495 'large' suitcases per day. Assuming the check-in service operates on every metro journey to the airport, this design could facilitate the movement of 1237 'large' suitcases per day. Despite this design being unable to accommodate as much baggage as Design 1 and 2.1, it still has the potential to transfer a substantial number of luggage per day while offering an alternative method of operating an on-board check-in and bag drop facility.

6.8.4 Design 3

Design 3 briefly looks at the possibility of combining features from Design 1, 2.1 and 2.2 to create a storage area of maximum capacity and a service which provides the greatest amount of facilities for passengers. This design shows the maximum potential of baggage that an on-board check-in and bag drop facility could operate for the Tyne and Wear Metro. The newly considered concept for this design is only considered within the storage area. The idea for this design would be to have the storage area for Design 1 incorporated onto all metro cars. If there is an expected passenger increase due to a large passenger flight departing or a high density of flights during a particular time of the day, the roll container could be incorporated with this design to temporarily increase the capacity of the service. This design also shows the potential to be able to modify further and develop the metro to increase its capacity if the on-board check-in and bag drop facility increases in popularity over time.

Check-in Desk

The check-in desk used would be the same design used in Design 2.1 and 2.2. This allows for the desk to be stored within the roll container and for the availability of printing boarding passes.

The Storage Area

For this design, all the facilities within the Design 1 storage area would be available apart from the double jointed drop down the arm, which acts as a barrier, which would be replaced as another single jointed 'fold-out arm' which contains a pull out strap within it. The extra arm would provide added support for safely stacking the luggage. For this design, the safety barrier used for Design 2.1 and 2.2 would be used.

Design 3 incorporates the roll container from Design 2.2 within the storage area while additionally containing the features from Design 1. Within Design 2.1 and 2.2, the seating space within the storage area is not occupied and therefore there is potential for this area to be accommodated for further storage for baggage.

The aim of this design is that luggage can be stored in the roll container while also being stored on the seats and on the luggage racks. The luggage racks are an instalment that would benefit all metro passengers and therefore should be utilised while the fold out arms to do not accommodate any capacity of the metro vehicle.

The roll container is transferred onto the metro and positioned and installed in the same way as in Design 2.2. The 'fold out arms' on the seating of the storage area remain untouched. This would complete the instalment of the storage area.

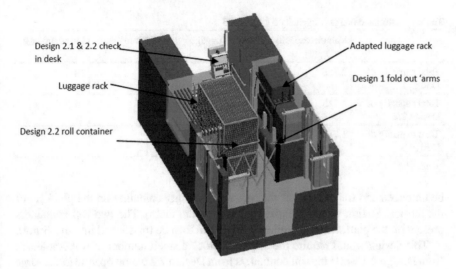

Fig. 55 Design 3 when in operation at full capacity

For this design to work, a specific order in the way the storage area and luggage are managed while the check-in desk is in operation is required. Firstly, luggage would be loaded onto the roll container as seen in Design 2.2.

Following the maximum capacity of the roll container being reached, luggage would be stored on the seating and luggage racks positioned on the opposite side to the roll container. The operation of this storage for the seats and luggage rack would be the same as seen in Design 1 however with the addition of an extra arm to raise up and a strap to connect to the luggage rack.

If further storage is still needed to store more baggage, it would be possible to unlock the roll container from the hand rails and push it to sit against the opposite longitudinal seats. This would prevent access to the luggage already stored on the seats, luggage rack and front access to the roll container; however, this luggage should not be accessible once checked in. Upon transferring the roll container to the opposite side, luggage can then be stored on the available seats and luggage rack exposed. This design of operation can be seen in Fig. 55. For this design to operate, it would be highly recommended for at least 2 members of staff to be on-board the metro at all times.

Upon arriving at the airport, two Newcastle International Airport staff would be waiting with two roll containers. One roll container would be of the design of the container seen in Design 2.1 while the other would be of the design from 2.2. Luggage stored on the metro on the seating and baggage racks would be transferred from the metro, onto the platform and into the roll container from Design 2.1. This roll container was found to have a luggage capacity equal to Design 1, and therefore it would be guaranteed that the entire luggage from the seating and baggage racks could be loaded onto the roll container. The container on-board the metro would then

Table 6 Estimated baggage capacity for Design 3

	Volume capacity (m^3)	Small luggage	Medium luggage	Large luggage
Total capacity of Design 1	2.358	46	30	18
Total capacity of Design 2.2	1.713	35	22	15
Total capacity of Design 3	1.713	81	52	33

be unlocked and transferred off the metro. The empty container on the platform of the Airport Station would then be loaded onto the metro. The two roll containers present on the platform, containing luggage, can then be transferred into the airport.

This design would require the airport to stock a small number of roll containers from Design 2.1 while the roll containers from Design 2.2 would operate in the same way.

Following the luggage being off-loaded from the metro vehicle and the new empty container being transferred on-board, the whole of the check-in and bag drop area would be stored away in the same way as seen in Design 2.2.

This design aims to show the how much baggage an on-board check-in and bag drop facility could accommodate and therefore provide an idea on the potential of such a service and show the possibilities in which a facility like this could be developed.

To calculate the luggage capacity of this Design, the values calculated from Design 1 and 2.2 were combined. These values can be seen in Table 6.

Moreover, again considering the worst case scenario in which all passengers' board with large luggage, this would give the facility a baggage capacity of 33 bags per journey—a capacity over twice that of Design 2.2. This system has the potential to facilitate the movement of 1089 'large' suitcases per day. Assuming the check-in service operates on every metro journey to the airport, this design could facilitate the movement of 2722 'large' suitcases per day.

This design would be a more expensive option; however it still delivers the key features and specifications required for the check-in and bag drop facility to be a success while producing an impressive volume of baggage movement. This shows there is great potential for this facility to operate at a remarkable scale while still only occupying a small area of the metro and allowing for a similar level of operation for the Tyne and Wear Metro.

6.9 Staff

For Design 1, 2.1, 2.2 and 3 to operate, at least one member of staff is required. This would be the cheapest option; however, for efficient operation of this service, two members of staff are recommended. An on-board check-in and bag drop facility is a brand new concept which has not been implemented anywhere in the world. The functionality of the service may, therefore, be confusing to many, particularly during the early days of opening. It is recommended that one staff member would operate behind the desk while the other would operate outside of it and manage the passengers. This would allow a staff member to provide clear instructions to passengers and aid with sufficient organisation thus preventing queuing and more efficient service. The staff member outside the check-in desk can operate with a tablet device and provide live information to passengers regarding flight information. While the staff member behind the check-in desk is serving customers and handling luggage, the other member of staff can answer any queries passengers have and provide sound advice. Further to this, this staff member can assist in the handling of luggage and storing away of the check-in equipment during peak times.

The staff member outside of the check-in desk can also assist non-airport passengers. Based on customer service attributes between 2009 and 2012, customer service factors which are consistently rated highly are staff availability, train cleanliness and passenger behaviour (Nexus 2013). These factors can be addressed with the introduction of this member of staff. With currently, just an occupied driver on-board the Tyne and Wear Metro, there is a poor staff availability and therefore this additional staff member can provide greater availability to all passengers. With the presence of a second staff member, passenger behaviour is likely to improve due to the fear of being removed from the metro and reported to the authorities. The cleanliness of the train can also be addressed as a quick clean can be conducted by the staff member at both South Hylton and the Airport Station.

6.10 Security

Security is extremely important for this concept. While the installation of an X-ray machine is not practical, certain protocols and features can be implemented to account for this. Major assistance of the security for this concept would be to have a second member of staff that could provide additional surveillance to potential threats.

A quote from the Tyne and Wear public transport users group states, 'we believe safety and security on the Metro should be a major priority, with steps taken to improve this where possible. On-train staff presence is essential for security. We would strongly welcome proposals that would increase staffing for passenger assistance in passenger areas of trains and in stations' (Tyne and Wear Public Transport Users Group 2010). Without passengers being assured that the baggage is safe and, most importantly, the passengers themselves are safe, customers of the Tyne and Wear Metro would be deterred from using the on-board check-in and bag drop facility. By introducing two members of staff on-board the metro while the check-in desk is in operation, this provides physiological and psychological assurance for passengers regarding safety. Further to this, the staff member outside of the check-in desk would be able to provide improved vigilance to threats and security breaches.

Specific practices based around the security for light rail vehicles can be introduced into the operation of the Tyne and Wear Metro. These include the development of operations for agency staff and employees to implement such as screening of passengers by observation techniques (SPOT) (APTA Standards Development Program 2013) and baggage reconciliation techniques to prevent a disembarking passenger leaving baggage behind (Departmnet for Transport 2014).

Passengers can act as public surveillance in observing threats on-board the metro. A public awareness campaign of 'if you see something, say something' can be heavily incorporated into the metro vehicles to raise public awareness of indicators of terrorism and violent crime (APTA Standards Development Program 2013). This can be done by the use of posters fitted within the metro vehicle and on the station platforms. If a passenger develops suspicion, a member of staff would be available on-board to then deal with this concern in a proper manner.

Currently, CCTV cameras are positioned by each set of doors on-board the metro. This allows for clear capture of every passenger that boards the metro. For these designs, a CCTV camera is recommended to be positioned to face the check-in desk and the storage area of the luggage. This provides constant surveillance of the luggage and check-in area at all times. These cameras provide a valuable source of evidence for use in the detection and investigation of crime thus act as a useful deterrent value (Departmnet for Transport 2014).

PSIs are believed to both deter and detect terrorist activity and are suspicion-less inspections of transit passengers by staff members (APTA Standards Development Program 2013). This technique is a quick process in which passengers can be randomly searched. This may involve luggage being randomly hand search after every set amount of luggage that has been checked in. The unpredictability of whether luggage will be searched is an effective deterrent for terrorist activity while also not producing a major change to the efficiency of the check-in service.

Attempting to implement on-board security equipment which airports facilitate, such as X-ray machines, to provide sufficient security checks to luggage before being loaded onto the aeroplane would be impossible due to the amount of power that would be consumed by the metro, the significant additional weight, the large capacity of the metro it would require and the predicable extensive reduction in the efficiency of the check-in facility. Complications would arise in transferring luggage directly from the metro onto the aeroplane, as sufficient security checks would want to be conducted by the airport rather than leaving it accountable to the Tyne and Wear Metro. Therefore, an effective way of assuring the luggage checked in on the metro undergoes sufficient security checks before being loaded onto the aeroplane is to simply transfer the luggage from the platform and into the terminal where the baggage can then be loaded via the bag drop desk utilised for extra-large baggage.

This bag drop area is used as a neutral dropping point for awkwardly sized and extra-large which has already been checked in and thus contains a baggage label, similarly to the baggage checked in on the metro. Luggage that is passed through here undergoes all the required security checks of the airport and is arranged into the corresponding destination of flight. By operating the on-board check-in and bag drop service such that each metro serves a set number of flights, this method of transferring the luggage through airport security has the potential to operate successfully. Once the entire luggage checked in on the metro is passed through the airport bag drop desk, the usual security checks and handling of the luggage can be implemented thus providing the same assurance for the safety of passengers on-board the flight.

7 Evaluation

7.1 Design Evaluation

By weighting the design criteria based on the importance of each criterion that contributes to the design, an overall weighted score could be produced to provide a ranking of the preferred designs for this study. The design criteria and weighting were developed through observations made during the study, review of the research and literature and discussions with experts from Nexus. Each design has been produced to meet the criteria; however, to obtain the optimum design, the ratings produced are relative to each other. Ratings follow a scale of 5 (very good) to 1 (very poor). The results for this evaluation are shown in Table 7.

Based on this evaluation, the optimum design which best considers the criteria set at the start of this study have been determined as Design 2.1. This was due to its simplistic design which allowed for exceptional performance, usability, robustness and aesthetics. Design 3 provides the impressive potential for the number of luggage it could accommodate in a single journey, however, this evaluation table highlights that in order for Design 3 to successfully operate, further considerations must be made to comply with the design criteria for safety and cost. A brief discussion behind the reasoning for the ratings given for each criterion can be seen below.

Performance—Design 2.1 scored the best for this criterion. This was due to the ability to be able to store a high volume of luggage while providing the ability for passengers to print boarding passes. This design can be operationally turned around at the airport in the most efficient and simplistic way and with minimal handling of equipment. This design exceeded that of Design 2.2 and three due to the operation being the simplest and having the ability to be installed and packed away efficiently. Design 1 storage area provided high capacity for the storage for luggage while providing sufficient floor space for the staff to work, however, due to the more complex setup of the check-in desk, it was deemed below the performance Design 2.1 provides.

Table 7 A table displaying the weighted scores for the design criteria for each design

Criteria	Weight (%)	Design 1		Design 2.1		Design 2.2		Design 3	
		Rating	Weighted score	Rating	Weighted score	Rating	Weighted score	Rating	Weighted score
Performance	14.7	4	0.588	5	0.735	4	0.588	4	0.588
Cost	14.7	3	0.441	3.5	0.5145	5	0.735	2	0.294
Versatility	14.7	5	0.735	4	0.588	5	0.735	4	0.588
Safety	14.7	5	0.735	5	0.735	5	0.735	3	0.441
Security	14.7	4	0.588	5	0.735	5	0.735	4	0.588
Usability	11.8	4	0.472	5	0.59	3.5	0.413	3	0.354
Robustness	8.8	2.5	0.22	4	0.352	3	0.264	3	0.264
Aesthetics	5.9	3.5	0.2065	4	0.236	4	0.236	3	0.177
Total weighted score		3.99		4.49		4.44		3.29	
Rank		3		1		2		4	

Cost—Design 2.2 scored the best for this criterion. This design required small permanent upgrades to the interior of the current metro vehicles. This avoids the need for all the vehicles to undergo extensive modifications. Instead, a specified amount of the equipment can be bought in and adapted to each metro as required. The check-in is composed of fewer components as the MD1 Service Desk incorporates most of the required facilities. A bulk order of one component from the same company results in cheaper costs overall. Design 2.1 required separate transportation for the transfer of the roll containers, which adds further employment and expense. Design 1 requires the purchase and manufacturing of several individual components, including an internal battery for each check-in desk, the addition of two fire extinguishers and new designs manufactured specifically for the metro in the case of the check-in desk, luggage racks and 'fold out arm' supports. Design 3 combines features from Design 1 and 2.1 and is, therefore, the most expensive.

Versatility—Design 2.2 and Design 1 scored best for this criterion. Compared with Design 2.1, 2.2 can retain the exact operation of the metro, by transferring the roll container back to South Hylton via the metro, while occupying no seating space. While design 1 limits the number of passengers it can provide a service to due to the inability to obtain a boarding pass on-board, this design provides the best design in which the smallest area is permanently accommodated and thus minimises the change of the internal operation of the metro and its capacity. Passengers can pass internally through the metro vehicle from side A to side B, and this allows for complete access and the same internal operation. By the utilisation of the internal battery, installation of plugs is avoided. Design 3 provides a similar operation as Design 2.2. However, an additional operation of roll containers within the airport is required.

Safety—All designs ensure that components are locked down during transit. This removes the potential for equipment to be projected down the aisles in the event of emergency braking or an impact. All designs ensure wires are encased by an enclosure for check-in desk equipment. The utilisation of the waiting area removes luggage from being a tripping hazard for non-airport passengers and more efficient boarding reducing the chance of crushing. Each design considers the availability of fire extinguishers is remaining available at all times. Design 3 scores lower due to the need to transfer a full capacity roll container from one side of the storage area to the other. This involves temporarily unlocking the container. Therefore, comprehensive communication between the driver and the staff member operating the desk must be implemented to prevent the metro departing while the container is unlocked. The roll container used for Design 2.1 minimised the number of time the luggage is moved by the train staff and prevents over overexertion.

Security—All designs incorporate storage areas equipped with locks to contain expensive equipment within the desk. Security procedures and equipment (see 5.7) would be implemented for all designs. The security of Design 2.1 and 2.2 exceeds that of Design 1 and three however as luggage is locked and stored away in an enclosed container during transit making unauthorised access to luggage highly improbable. Furthermore, the safety barrier used would provide more extensive prevention into the storage area due to the larger area it barricades and greater difficulty in dissembling.

Usability—Design 2.1 provides the simplest operation in terms of storing the equipment away. Handling of the luggage is minimal while the check-in desk does not require to be repositioned. This gives for a quick turnaround time. The MD1 Service Desk offers the widest range of facilities while providing an LCD screen for passengers thus making the facility more 'user-friendly'. Design 2.1 and 3 require the check-in desk and barrier to be detached and loaded onto a roll container which is then locked into position. This creates a more complicated storing process and is a more strenuous task. Design 3 scores worse here, however due to the requirement of staff managing more extensive and complex operation of the storage area. Design 1 induces a simple setup of the storage area; however the installation and packing away of the check-in desk requires a more convoluted process than Design 2.1. However, Design 1 scores higher than Design 2.2 and 3 as the check-in desk for this design is a more compact and light weight design thus making manoeuvring it more easier.

Robustness—while this area will be considered more thoroughly following the production of prototypes of the proposed designs; an initial estimation can be made as to the robustness of the designs. Design 1 scores low relative to the other designs due to the frequency of demand and use of the components; constantly building and deconstructing the check-in desk can lead to wear over time. The more components, the greater chance there is of damage to the facility. The roll top mechanism could be susceptible to grit and dirt contamination that would cause damage over time. Design 2.1 scores the highest in this criterion as the equipment is subject to the least handling and therefore less likely to be damaged. The check-in desk used for Design 2.1, 2.2 and 3 contains a casing for the weighing scale and thereby removes the need to handle the weighing scale. The robustness rating for Design 2.2 and 3 is lower than that of 2.1 due to the drop down ramp within rolling container which can be compromised by regular heavy loads thus inducing repetitive strain on the hinges of the ramp.

Aesthetics—The check-in desk of Design 2.1, 2.2 and 3 sits within the width of the aisle while being composed of only one part. This gives a more complete, sturdy and professional finish to the product. With the MD1 Service Desk providing LCD screens for the passengers, this provides a modern and technologically advanced service which would appeal to the modern passengers. The storage area of Design 1 provides a more subtle facility and gives a neater finish with floor space being retained. Despite this benefit of Design 1, the passengers will be more enticed by the aesthetics of the waiting area and check-in desk utilised by passengers and therefore Design 2.1 and 2.2 were scored higher. Design 3 was scored the lowest as it contains a very packed storage area which may not look aesthetically pleasing. Further to prototypes being produced, the aesthetics of each proposed design can be modified as this was not prioritised for this study.

7.2 Customer Opinion

Through research of a National Rail Passenger Survey (RSSB 2016), the key factors which influence customer satisfaction on a rail vehicle were observed. Factors which were applicable in terms of them being affected by the on-board check-in and bag drop facility were evaluated and scored based on the influence the designs and operations proposed would have on each area of customer satisfaction relating to passengers travelling to the airport. The scoring system used can be seen in Table 8.

The averages of the scores were deduced for each design. The results of this evaluation can be seen in Table 9.

Table 9 highlights the positive consideration each design has towards customer satisfaction. There is an extensive range of customer satisfaction considerations that are improved for airport passengers by the installation of this facility. This evaluation, therefore, shows that the end user would be excited by the designs proposed and opportunities provided. Design 2.1 and 2.2 would provide a service that would provide the highest customer satisfaction, however, due to similar operations for the internal designs being similar, the choice of design should not heavily impact the customer satisfaction, providing the operations proposed in this study are followed.

Passengers always able to get a seat on the train—Sufficient room being available for passengers to sit/stand is highlighted as currently one of the biggest impacts on overall dissatisfaction (RSSB 2016). For airport-bound passengers, this satisfaction is positively affected for all designs as the waiting area would provide priority seating. This waiting area provides six seats; however, this would not be enough to guarantee airport passengers a seat, particularly during busy periods. The designs do provide a move toward improving this consideration. Furthermore, once the luggage is checked in, access to other seating down the narrow aisles of the carriages would be more accessible.

Journey time reduced—By combining the check-in process and the journey to the airport into one transaction, the overall time a person would spend before passing through security would be reduced.

Punctuality/Reliability—The punctuality and reliability of rail service is the biggest impact on the overall satisfaction of the public transport. It is therefore important that this satisfaction is retained to not disrupt the positive reputation of rail services. This is maintained by ensuring there are no delays due to the check-in

Table 8 Scoring system used for the customer opinion evaluation

Customer satisfaction scoring system	
1	Highly positively effected
2	Positively effected
3	No effect
4	Negatively effected
5	Highly negatively effected

Table 9 Results of evaluation regarding the customer satisfaction obtained from the proposed designs

Customer satisfaction	Design 1	Design 2.1	Design 2.2	Design 3
Passengers always able to get a seat on the train	4	4	4	4
Journey time reduced	4	4	4	4
Punctuality/Reliability	4	4	4	2
Inside of train is maintained and cleaned to high standard	4	4	4	3
Accurate and timely information provided on trains	5	5	5	5
Personal security on trains	4	4	4	4
Sufficient space on train for passenger luggage	5	5	5	5
More staff available on trains to help passengers	5	5	5	5
Ease of getting on and off	4	4	4	4
Navigating the station	4	4	4	4
Boarding quickly and easily	5	5	5	5
High level of technology expected	4	5	5	5
Clear gangways	5	5	5	5
Total	**4.38**	**4.46**	**4.46**	**4.23**

and bag drop facility. To aid in this, the equipment must be able to be installed and packed away within the current operative time scale. Design 1, 2.1 and 2.2 could be installed and packed away within the 10 min turnaround time at the Airport and South Hylton Station. This provides the most reliable transport to the airport when compared to road transportation. Design 3 would require additional time to handle the extensive amount of equipment and operations contained within it and thus reduce the punctuality and regularity of the metro.

Inside of train is maintained and cleaned—With the addition of extra staff on-board the metro service, there is an opportunity for staff to conduct a quick clean of the vehicle during turnaround when arriving at the Airport and South Hylton Station. Design 3 would find difficulty in adding this requirement to the already extensive tasks required from the staff on-board at turnaround.

Accurate and timely information provided on trains—With a member of staff operating outside of the check-in area, this staff member will provide live information through the use of a smartphone/tablet device.

Personal security on trains—There will always be concern from some passengers regarding the storage of luggage on trains. Each design has considered the optimal solution to maintain safety and assure passengers. The additional staff provided also helps with personal security.

Sufficient space on the train for passenger luggage—With airport passengers being able to check-in and drop off luggage on-board the metro, this customer satisfaction area has been completely resolved.

More staff available on trains to help passengers—The purpose of the extra staff member operating outside of the check-in area is to assist passengers. This is a feature which currently is not present on the Tyne and Wear Metro.

Ease of getting on and off—Departing the metro at the airport station is easier with no luggage to have to carry. There remains a difficulty for passengers boarding the metro when travelling to the airport and when boarding and departing on the return journey from the airport. These designs provide some improvement to this area of satisfaction.

Navigating the Station—This study considered the need for clear signs and markings to ensure airport passengers understand where to position themselves and where to board the metro.

Boarding quickly and easily—A smoother boarding process would be implemented from these designs as passengers with luggage would have clear access to the waiting area. This removes the chance of a bottle neck effect down the narrow aisles caused by passengers with luggage attempting to manoeuver between the seats and taking up space by placing luggage on seats and the floor; all of which can prevent access for other passengers.

High level of technology expected—The presence of an on-board check-in and bag drop facility would be a completely new concept and the presence of this would incorporate a high level of technology. The MD1 Service Desk provides a greater range of facilities than in Design 1; including LCD screens for the passengers.

Clear gangways—As in the Tyne and Wear Metro, the lack of storage causes seats and aisles to be taken up by baggage. This reduces the capacity of the metro, creates a safety hazard and reduces the efficient flow of passengers inside the metro. The storage area safely stores away any airport related baggage on-board the metro and clears up the gangways.

7.3 Measure of System Performance Comparison with Existing Solutions

To provide a valid assessment of the proposed design and operations of the on-board check-in system, comparisons have been made against existing systems and performance. This section of the evaluation provides a clear comparison with the differing performances available with a selection of alternative check-in and bag drop facilities to highlight the benefits and limitations of the design proposed from this study. To evaluate this study, Design 2.1 has been used to compare with the existing systems as this was deemed the best design through comparison with the other proposed designs and customer opinion. This assessment can be seen in Table 10.

Overall, this evaluation shows that the on-board check-in and bag drop facility can provide similar features to the already successful Hong Kong in town check in. Design 2.1 may be forced to impose stricter rules surrounding luggage size and weight, however, it provides a versatile opportunity which can be adapted to existing rail services and thus reduces the need for change to the infrastructure and therefore cost. Furthermore, the design provides a wider range of locations where passengers can check-in luggage is therefore making it accessible and convenient for a greater percentage of the population. A further look at the measures of system performances is discussed next.

Baggage Capacity—Design 2.1 would limit the maximum luggage that could be transferred per day, unlike the other systems in place. The design should, however, provide a sufficient capacity of luggage for Newcastle International Airport with it being one terminal.

Maximum Luggage Size—This design would provide a smaller than average maximum luggage size. Alterations could easily be made to this by incorporating the same concept from Hong Kong in town check in where passengers simply pay extra to check-in larger luggage. This assures the system retains the same turn over for each journey made. The seating within the storage area could be used to store larger items of luggage.

Maximum Luggage Weight—For Design 2.1, a strict 30 kg luggage allowance should be enforced. Luggage loaded onto the roll container is then required to be pushed into the airport; baggage which exceeds the 30 kg weight would make it impossible and unsafe for the member of staff to navigate the container from the metro and into the airport.

Number of Bag Drop Locations—The proposed on-board check-in and bag drop facility offers a bag drop location nearby for most of the population in the Tyne and Wear area. The Hong Kong and Newcastle City Centre bag drop limit the location to which passengers can check-in luggage and create an added inconvenience.

Luggage Collection Locations—Like the Hong Kong check-in, Design 2.1 allows for luggage to be collected in the usual way at the baggage collection point at the airport destination.

Time Taken for Luggage to Arrive at Destination—By allowing the handling of the luggage to be handed over to the airport staff and airport processes followed, the Check In and Bag Drop Service ensures that the usual time taken for luggage to arrive at the passenger's destination is retained and means that the luggage flies with the passenger. Virgin Bag Magic provides interesting benefits however is greatly limited by the inability to receive the luggage upon arriving at the destination.

Time Passenger is Baggageless—Passengers using the Hong Kong and Newcastle bag drop facility must potentially use extensive public transport to arrive at the in

Table 10 A comparison of Design 2.1 with existing systems using measures of system performance

Measures of system performance	Design 2.1	Hong Kong in-town check-in (InPost 2016; RSSB 2016)	Daniel Brice's Newcastle bag drop (Reece and Marinov (2015b)	Virgin Bag Magic (Baker 2012)
Baggage capacity of service	Up to 1485 'large' baggage per day	Unlimited	Every single passenger on every plane could be accommodated for	Unlimited
Maximum Luggage Size	Large suitcase (0.11 m^3)	Unlimited (bags that exceed 0.209 m^3 induce added cost)	Unlimited (costing not discussed for this study)	0.189 m^3
Maximum Luggage Weight	30 kg	Unlimited (bags that exceed 30 kg induce added cost)	Unlimited (costing not discussed for this study)	30 kg
Number of bag drop locations	30 (bags can be checked in by boarding an airport metro at any of the 30 stations prior to the airport)	2 (Hong Kong Station or Kowloon Station)	1 (Haymarket Station in Newcastle Centre)	Unlimited (bags collected from desired location)
Luggage collection location	Collected from luggage collection point at airport destination	Collected from luggage collection point at airport destination	Area of study not discussed yet	Luggage dropped off at requested address
Time taken for luggage to arrive at destination	Dependant on the flight time and airport destination handling times	Dependant on the flight time and airport destination handling times	Area of study not discussed yet	Over 24 h
Time passenger is baggageless	Passenger become baggageless soon after boarding the metro	Passenger becomes baggageless following arriving at Hong Kong/Kowloon Station	Passenger becomes baggageless following arriving at Haymarket Station (Newcastle City Centre)	Passenger baggageless for whole journey

(continued)

Table 10 (continued)

Measures of system performance	Design 2.1	Hong Kong in-town check-in (InPost 2016; RSSB 2016)	Daniel Brice's Newcastle bag drop (Reece and Marinov (2015b)	Virgin Bag Magic (Baker 2012)
Additional preparation compared to standard airport check-in	Purchase of airport metro check-in ticket	Purchase of Airport Express train ticket	Area of study not discussed yet	Service must be paid for online and luggage must be packed a day in advance
Infrastructure adaptation and construction required	No change to infrastructure required	New rail track constructed specifically for check- in area. Construction of station required with this	Extension of Haymarket Station would be required	No change to infrastructure required (existing road courier service used to transport bags)
Hours of operation	6 a.m. to 12 p.m. (18 h)	5.30 a.m. to 90 min before last flight departure	Freight train operates 24 h per day	Collection and delivery operates Mon–Fri, 9 a.m. to 5 p.m.
Destination availability	Any destination	Any destination	Any destination	Within the UK

town check-in desk. Design 2.1 provides a service where passengers can travel to the nearest metro station to check-in baggage.

Additional Preparation Compared to Standard Airport Check-In—Virgin Bag Magic requires luggage to be prepacked and collected well in advance of the flight making it unsuitable for last minute flights. Just like the Hong Kong check in, passengers can use the facility close to the departure of the flight.

Infrastructure Adaptation and Construction Required—A primary benefit with the proposed design is the requirement not to change the current infrastructure of the Tyne and Wear Metro. The Newcastle bag drop proposal would require major changes to Haymarket Station and the operation of the metro and thus making the funding for this unrealistic. In the case of the Hong Kong check-in, this required new stations and rail tracks where such funding for this would not be available for the Tyne and Wear area.

Hours of Operation—The metro currently already operates within a similar schedule to flight departures from Newcastle International Airport. Therefore the metro would be available when most flights are departing. To improve this design further however, the metro would need to operate to a timescale similar to the Hong Kong in town check-in. Virgin Bag Magic is limited as it does not operate outside of working times.

Destination Availability—With the design and operation proposed, it is easy to accommodate all passengers no matter what the destination of luggage is. Handling of the luggage by the Tyne and Wear Metro ceases at the Airport Station, and therefore the destination of the luggage does not need to be considered. This applies to other existing systems however Virgin Bag Magic handles the luggage for the whole process and therefore are limited to delivery inside the UK only.

8 Technical Validation

To validate the design, assurance is needed to provide certainty that the metro can still operate with the added weight of the facilities and storage of luggage. As quoted by Nexus, the passenger capacity of a Tyne and Wear Metro car is assumed to be six passengers per square metre standing. The storage area would be considered as the area containing the highest increase in additional mass. It has a standing area of 2.53 m^2 and accommodates six seats. This considered the maximum capacity of the storage area is 15 standing passengers plus six seated passengers. The totals are to the metro for being able to operate with 21 passengers in the storage area. Considering 21 persons weighing 80 kg, the maximum weight capacity could accumulate to over 1.6 tonnes. In the case of Design 3, the design with the greatest additional mass, the storage area would be required to hold 33 items of luggage weighing a maximum of 30 kg each. This would accumulate to a mass of 990 kg. This would leave a further 600 kg of weight which the metro axles could support. With just the further addition of a roll container, safety barrier, luggage racks and 'fold out arms', it is not possible for these pieces of equipment to exceed 600 kg and therefore it can be said that the axle loading is acceptable.

Features based on the 'Initial Design Limitations' have been applied to all designs to allow for a valid design solution to be adapted into the Tyne and Wear Metro and assure the designs follow all the requirements currently in place on the metro vehicles. These features can be seen in the design solutions for Design 1, 2.1, 2.2 and 3.

9 Conclusions and Future Work

This study presents solutions to the designs and operations required to have an on-board check-in and bag drop facility on a light rail service. Conclusions can be drawn about the possibility of such a facility being enabled.

There are two potential areas for a check-in desk in a metro vehicle. These are: at the ends and at the centre of the vehicle. This study concluded that the centre of the Tyne and Wear Metro carriage provided the optimum area due to its pre-existing layout which provided enough floor space, by use of longitudinal seating and wider aisles, which suited towards passengers with luggage. This open space provided enough room to install equipment without the need to reconfigure the internal layout of the vehicle. The pivoting tractor bogie located beneath the centre of most light rail vehicles, provides a layout consistent with other light rail vehicles. A low seating density and absence of doors in this area creates a heavily pre-fabricated area.

Cost funding is an influential factor in determining the positioning of the desk. While positioning the desk at the ends of the metro vehicle might accommodate more luggage and wider range of facilities, it would actually require extensive internal redesign and relocating of the current equipment. The engineering involved would certainly cost more.

Good organisation is vital if this proposal is to become a reality. Realistic terms must be met in which the check-in service on each metro vehicle is provided to a specific set of passengers destined to a predetermined destination. If each metro vehicle operated for specific flights, estimated numbers of passengers can be predicted for each metro journey, and equipment programmed to check-in for a set number of flights. This would control the passenger flow and improve the organisation of the luggage. Clear communication between the airport and the metro service is essential to provide an efficient service without affecting the punctuality of the metro system. Retaining customer satisfaction amongst all metro users is vital. If regular riders are deterred, the overall passenger may not increase.

Simplified details of how the facility operates can be shown to passengers by the development of smartphone applications, website information and signs on both the platform and interior of the metro vehicle. This study shows that a straightforward design can be developed which would provide little confusion amongst passengers.

The limitations uncovered during studying the Tyne and Wear Metro in Newcastle could also apply to other metro and light rail vehicles. To keep cost down, it is important that the current vehicle configuration is retained. Thus, compact designs have been proposed which do not use much of the vehicle. The check-in and bag drop facility should not be a permanent installation to an existing metro vehicle as this would generate unnecessary costs and occupying space on rail vehicles not travelling to the airport. In the case of the Tyne and Wear Metro, where not all vehicles go to the airport, the ability to remove and install equipment efficiently depending on the destination of the vehicle has been achieved by producing designs which are able to be transferred to other vehicles. This allows metro vehicles to operate in the usual way when not serving the airport.

Comprehensive health and safety factors are implemented in most metro and light rail vehicles around the world. A design which tampers with these factors makes it harder to gain acceptance of the check-in facility. The health and safety features on the rail vehicle should be maintained for any design developed. In the case of the Tyne and Wear Metro, emergency door handles, fire extinguishers and disabled access were all unaffected by the introduction of the check-in and bag drop facility.

Efficient boarding and organisation of passengers was achieved through the use of a waiting area which was possible due to the location of the check-in area. Priority seating is provided in this area which offers an incentive for airport passengers to use metro and rail transport. To help to fund this facility and to enlist the airport's support, advertising space could be offered within the waiting area, sponsored by the airport.

With each design produced for this study, safety and security have been seriously considered, both in the event of an emergency and during a routine journey. All designs are positioned and then locked down to the surrounding hand rails, thus assuring the stability of the desks. It was concluded that the installation of X-rays machines would not be possible, however, through the use of CCTV, public awareness and passenger inspections, a sufficient level of security could be achieved. Transferring luggage from the metro and straight onto the aeroplane was considered but this was thought unrealistic due to security issues where the airport cannot allow an external source to check luggage outside the airport building. Luggage would, therefore, be transported to a baggage drop point inside the airport terminal which allows all the usual security checks to be administered by the airport.

Design 1 provides a highly versatile design that can be adapted effectively to the current metro vehicle. The compact design of both the check-in desk and storage area allows a similar operation of the rail vehicle when the equipment is stored away. The storage area is a simple design using small components that can be easily set up and collapsed and take up no capacity on the metro when not in use. A luggage rack has been designed proving for available space for them to be installed whilst this facility can be applied to all metro vehicles and improve customer satisfaction amongst all passengers. The check-in desk is easily manoeuvrable which is vital as the desk is relocated after arrival at the airport station. To assist in a compact design and the security of the equipment, an extensive internal storage area has been designed into the desk where all the equipment from the on-board check-in and bag drop facility can be stored inside and safely locked away.

Design 2.1 provided the best outcome following the evaluation of the designs. The performance of the desk was optimised in providing comprehensive facilities for passengers whilst ensuring a large storage area. The use of a roll container removes the need to unload luggage from the metro vehicle, saving time and allowing for an efficient process. Security of the luggage is improved as baggage would remain in the container until inside the airport. When arriving at the airport, it is a quick process to convert the rail vehicle to reopen the area to all passengers. A negligible area of floor space is required permanently for the check-in desk when the facility is installed, but there is a cost for separate transportation of the roll containers to be returned to the first station of the rail line. For larger light rail services this may cause an issue, however, alternative designs within this study can be considered.

Design 2.2 removes the issue of requiring separate transportation of the roll containers. This is done through the implementation of a side ramp installed on the container which allows for the check-in desk to be stored inside. Thus, all equipment can be transported on the rail vehicle when travelling away from the airport, minimising the amount of room it uses. Damage of equipment is prevented as equipment can be stored inside the container and thus out of reach. Handling of the equipment would be harder for this design. However, the process remains simple.

Design 3 showed that in the case of the Tyne and Wear Metro, an impressive amount of storage could be accommodated within a small area. This setup shows that there is potential for high capacity luggage check-in facilities to be available on rail services even where there is limited room. Evaluation of this design showed increased costs, uncertainty about both the safety of the operation required as well as about the usability of the equipment. This design would require more staff members to operate it and take more amount of time to unload luggage and store away equipment which may result in the need for alterations to the current operation of the metro service.

The highest cost predicted for operating an on-board check-in desk on-board a metro vehicle is a staff required. Research from this study suggests that two members of staff should be on-board the vehicle where a check-in facility is used. Therefore implementing this for every metro vehicle containing the check-in desk may be expensive. The positive factor is that it creates employment.

The research into the customer perspective towards the designs proposed was positive and suggested that several key areas of customer satisfaction would be addressed which in turn would help generate interest in the use of the facility. This shows that there would be a desire for the on-board check-in and bag drop facility to be implemented on a metro service.

Comparisons of existing systems with the designs and operations proposed in this study suggest a similar service can be provided but without the need for change to the operation of the rail vehicles and without major reconstruction of track line and stations. Limitations are highlighted in the maximum capacity of luggage however, this can be resolved through the use of the check-in desk in Design 2.1 where a low working surface is designed in which large luggage could be passed over. Additional costs could then be charged for baggage that exceeds the size of a 'large' suitcase.

In conclusion, there is clear evidence that there is an opportunity for rail vehicles to have an on-board check-in and bag drop service. The designs produced have successfully met the design criteria and offered realistic and achievable interior designs and operations which can be further investigated.

There are many considerations in achieving a successful check-in service on a rail vehicle. This study has highlighted the key challenges presented and offers intuitive designs which allow for the functionality of the check-in facility. The study suggests the possibility for a check-in and bag drop service to be implemented without the need for large infrastructure changes in the cases of the redevelopment of stations and signalling systems in place. In the case of each design, the functionalities which airport check-in desks provide can be installed onto a rail vehicle whilst sufficient storage for luggage can be provided without the need to use a large amount of the vehicle capacity. It has been proven that vehicle modifications can be kept to a minimum and thus this concept can be applied to a variety of global rail services without the need for extensive funding and major redevelopment of existing rail vehicles.

A cost benefit analysis must be constructed for this study. This would provide value as to the estimated funding required to implement the designs and operations proposed. This can obtain the advantages and limitations raised and convert them into an analysis of whether the overall outcome would provide a beneficial service. Understanding the costings of the designs can assist in providing further comparisons with existing systems before then being able to estimate a cost at which the on-board bag drop and check-in service would be able to be provided for the customers. This would help in better understanding the additional number of staff that would be ideal for the proposed designs and operations. Analysis can be conducted which observes the most cost beneficial and effective service in terms of the ratio of metro vehicles travelling to the airport equipped with the check-in facility. Further development of these designs should be avoided until this analysis is conducted.

The operations and designs from this study should be generated into a computer simulation. This would allow for estimations to be calculated for the number of passengers likely to use the facility based on the efficiency and costing of the service. Different operations can be proposed as to how to accommodate the equipment and then assessed to determine the most effective operation. Values can be produced as to the estimated time each design would take to be installed and stored away which would provide a better understanding as to whether the designs are able to work within the current time scales and operation of the Tyne and Wear Metro. Conducting this simulation would provide a greater validity of the designs and operations proposed.

A customer survey should be conducted within the Tyne and Wear area to acquire feedback on the implementation of the on-board check-in and bag drop facility. A clear display of the operations proposed and the designs produced can be shown to Tyne and Wear passengers to acquire feedback as to the preference of designs and whether such a facility would be utilised by themselves and if it would positively or negatively affect the opinion of the rail service.

Further analysis of the designs proposed must be done by assessment of a material study of the components. By selecting the best materials to use, a better understanding of the additional weight added to the metro, the manoeuvrability of the equipment and the overall costing can be obtained. Analysis by finite element methods can be used to conduct in-depth structural analysis of the proposed designs to provide a better understanding as to the robustness of the components which can assist in a more conclusive evaluation.

Managing the external operations should be looked at in greater depth. This study contains brief discussions as to how the equipment would be required to operate outside of the metro, however, this should be investigated further to understand the facilities and storage Nexus and Newcastle International Airport could provide and how the equipment would be managed and transferred across different rail vehicles.

To investigate the maximum potential the Tyne and Wear Metro has, a similar study should be conducted in which the check-in desk is observed at the end of the metro. This would require a greater amount of reconstruction of the vehicle and costing, however, the potential for an increased capacity of luggage and wider range of equipment may raise interesting considerations as to how an on-board check-in and bag drop facility could operate on a rail service and should, therefore, be investigated to be compared with the proposals from this study.

Acknowledgements The Mechanical Engineering MSc at Newcastle University, North East England, because this innovative research has been initiated as an MSc project within the scope of this postgraduate taught programme. An interim report is available at: https://eprint.ncl.ac.uk/file_store/production/226664/AB922309-0A52-434C-910D-A6499BBF033A.pdf (accessed on 4th July 2018).

The contribution of David Mee, Principal Engineer of Nexus Rail, for providing expert opinions throughout the study and answering desired questions contributing to the research and designs proposed.

Data Availability Statement
No raw or source data were collected through frequent observations and used to support the developments and findings of this study. The data collected during discussions with Tyne and Wear Metro and the general public are currently under embargo while the research findings are possibly commercialised. Requests for data, 60 months after publication of this article, will be considered by the corresponding author.

Funding Statement
This research did not receive any specific funding. Instead, it was performed as part of the employment and educational activities of authors at Newcastle University, North East England.

References

Aditjandra PT, Galatioto F, Bell MC, Zunder TH (2016) Evaluating the impacts of urban freight traffic: application of micro-simulation at a large establishment. Eur J Transp Infrastruct Res 16(1):4–22

APTA Standards Development Program (2013) Security operations for publics transit. American Public Transportation Association, Washington

Baker N (2012) Transport Minister inspects Tyne & Wear metro progress. Rail.co.uk (Online). Available: http://www.rail.co.uk/rail-news/2012/transport-minister-inspects-tyne-wear-metro-progress/. Accessed June 2016

Benjelloun A, Crainic T (2008) Trends, challenges and perspectives in city logistics. TRANSPORTUL ŞI AMENAJAREA TERITORIULUI, Buletinul AGIR nr. 4/2009, octombrie–decembrie

Bernhard R (2017) Influence of passenger behaviour on railway-station infrastructure, sustainable rail transport. In Fraszczyk A, Marinov M (eds) Proceedings of RailNewcastle 2017

Brice D, Marinov M (2015) A newly designed baggage transfer system implemented using event-based simulations. Urban Rail Transit 1(4):194–214

Civil Aviation Authority (2015) Main Outputs of Reporting Airports 1991–2015. Heathrow (Online). Available: https://www.caa.co.uk/uploadedFiles/CAA/Content/Standard_Content/Data_and_analysis/Datasets/Airport_stats/Airport_data_2015/Table_02_1_Main_Outputs_Of_UK_Airports_2015.pdf. Accessed 2016

Dablanc L (2007) Goods transport in large European cities: difficult to organize, difficult to modernize. Transp Res Part A: Policy Pract 41(3):280–285

Dampier A, Marinov M (2015) A study of the feasibility and potential implementation of metro-based freight transportation in Newcastle upon Tyne. Urban Rail Transit 1(3):164–182

Darlton A, Marinov M (2015) Suitability of tilting technology to the Tyne and Wear Metro System. Urban Rail Transit 1(1):47–68

Department for Transport (2014) Light rail security recommended best practice. London

Department for Transport (2015) Transport Statistics Great Britain 2015 (Online). Available: https://www.gov.uk/government/uploads/system/uploads/attachment_data/file/489894/tsgb-2015.pdf. Accessed July 2016

Department for Transport (2016) Road Traffic Estimates: Great Britain 2015, 19 May 2016 (Online). Available: https://www.gov.uk/government/uploads/system/uploads/attachment_data/file/524261/annual-road-traffic-estimates-2015.pdf. Accessed June 2016

Desko (2016) Desko Xenon 1900 Brochure (Online). Available: https://www.desko.de/cms-wAssets/docs/brochures/Brochure_DESKO_Xenon_1900.pdf. Accessed June 2016

Embross (2016) MD1 Mobile service desk. Embross (Online). Available: http://www.embross.com/md1-mobile-desk/. Accessed July 2016

GOV.UK (2011) Public experiences of and attitudes towards air travel (Online). Available: https://www.gov.uk/government/uploads/system/uploads/attachment_data/file/8907/attitudes-towards-air-travel-feb-2010.pdf. Accessed 2016

Harari D (2016) Regional and local economic growth stastics. House of Commons Library

Henley W (2013) Is it greener to travel by rail or car?. The Guardian, UK

InPost (2016) About InPost (Online). Available: https://inpost.co.uk/. Accessed May 2016

Jaffe E (2014) Every city needs Hong Kong's brilliant baggage-check system. City Lab, 22 August 2014 (Online). Available: http://www.citylab.com/cityfixer/2014/08/every-city-needs-hong-kongs-brilliant-baggage-check-system/378826/. Accessed June 2016

Jaller M, Wang X (Cara), Holguín-Veras J (2015) Large urban freight traffic generators: opportunities for city logistics initiatives. J Transp Land Use 8(1):1–17

Kelly J, Marinov M (2017) Innovative interior designs for urban freight distribution using light rail systems. Urban Rail Transit 2017, epub ahead of print

Lewis H (2012) Building the Tyne and Wear Metro. The Chartered Institution of Highways & Transportation, Newcastle Upon Tyne

Marinov M, Giubilei F, Gerhardt M, Özkan T, Stergiou E, Papadopol M, Cabecinha L (2013) Urban freight movement by rail. J Transp Lit 7(3):87–116

Metro (2016) Metro maps. Nexus (Online). Available: http://www.nexus.org.uk/metro/metro-maps. Accessed 2016

Monorails Australia (2014) Melbourne Airport Monorail, November 2014 (Online). Available: http://www.monorailsaustralia.com.au/airport.html. Accessed May 2016

Motraghi A, Marinov M (2012) Analysis of urban freight by rail using event based simulation. Simul Modell Pract Theor 25:73–89

Nexus (2013) Metro Strategy 2030—Background Information. Newcastle Upon Tyne

Nexus (2014) Metro timetable—South Hylton, January 2014 (Online). Available: http://www.nexus.org.uk/sites/default/files/metro/stations/MT1312.A4.SHL_.pdf. Accessed Aug 2016

Office for National Statistics (2012) Census result shows increase in population of the North East. News Release

Ogden KW (1977) An analysis of urban commodity flow. Transp Plann Technol 4(1):1–9

Passenger Statistics (2016) Newcastle International Airport, 2016 (Online). Available: http://www.newcastleairport.com/passengerstatistics. Accessed Feb 2016

Reece DC, Marinov M (2015a) How to facilitate the movement of passengers by introducing baggage collection systems for travel from North Shields to Newcastle International Airport. Transp Probl 1(1)

Reece DC, Marinov M (2015b) Modelling the implementation of a baggage transport system in Newcastle Upon Tyne for passengers using mixed-mode travel. Transp Probl 1(1)

RSSB (2016) Faster, safer, better: boarding and alighting trains. Rail Research UK Association

Rüger DB (2016) Baggageless. Vienna University of Technology- netwiss (Online). Available: http://publik.tuwien.ac.at/files/PubDat_232400.pdf. Accessed 2016

RVAR (2015) RVAR and PRM TSI Compliance, March 2015 (Online). Available: https://webcache.googleusercontent.com/search?q=cache:5kaCsILEA40J:https://www.gov.uk/government/uploads/system/uploads/attachment_data/file/426131/class-365-erg-compliance-assessment.xls+&cd=1&hl=en&ct=clnk&gl=uk. Accessed Aug 2016

Savrasovsa M, Medvedev A, Sincov E (2007) Riga airport baggage handling system simulation. Transport and Telecommunication Institute, Riga, Latvia

Sita.Aero (2016) Heavy duty printers for major airports (Online). Available: https://sita.aero/globalassets/microsites/atis-2016/sponsor-brochures/custom-atis2016.pdf. Accessed July 2016

Tyne and Wear Public Transport Users Group (2015) Metro strategy 2030—Response from Tyne and Wear Public (Online). Available: http://www.twptug.org.uk/download/i/mark_dl/u/4009947148/4613675771/Metro%20Consultation%20v6.pdf. Accessed 2016

Univeristy of Cambridge (2010) Technology roadmapping: facilitating collaborative strategy development. IfM Management Technology Policy, Cambridge

Wales J, Marinov M (2015) Analysis of delays and delay mitigation on a metropolitan railway network using event based simulation. Simul Modell Pract Theor 52:55–77

Weigh-Tronix A (2015) H305 low profile bench scale (Online). Available: http://www.averyweigh-tronix.nl/wp-content/uploads/2015/04/H305Weegplateau.pdf. Accessed June 2016

Workplace Products (2016) Titan expanding barriers. Workplace products (Online). Available: http://www.workplace-products.co.uk/titan-expanding-barriers-3-metre.html. Accessed July 2016

Woudsma C (2001) Understanding the movement of goods, not people: issues, evidence and potential. Urban Stud 38(13):2439–2455

Quality Assessment of Regional Railway Passenger Transport

Borna Abramović and Denis Šipuš

Abstract The passenger transport sector in the European Union today is very customer-oriented. Transport operators have recognized that having a satisfied customer means a passenger will return to use the service again and so increase the overall income of the transport operator. In railway passenger transport over the last decade, there have been efforts to place the passenger in the focus of the railway transport process, but the overall effort has been very slow. The primary focus is usually on railway lines with great demand, especially high-speed train lines and national train lines. Various methodologies that support this effort are different methods of quality of transport services. Quality is a fundamental influence on services in passenger transport. One of the most useful methods for assessing the quality of service is surveys. Surveys can be done from the passenger point of view but also from the transport operator's point of view. During the past decade the level of service on regional railway lines in Croatia is relatively stable with a trend towards downgrading the level of service. One of the observations is there is no regular assessment of services and no interaction with passengers. This paper presents the results of a survey-based quality assessment of passenger transport services on a regional line in the north-western part of Croatia where a comprehensive survey identified how to maintain the existing level of service, some proposals on how to improve the level of service are also presented.

Keywords Quality of service · Railway · Survey · Passengers · Regional

1 Introduction

A hot topic in transport over the last few years has been the mobility of the population. The latest White Paper Roadmap to a Single European Transport Area - Towards a competitive and resource efficient transport system (European Commission 2011) specifically emphasised the development of mobility as a service. The

B. Abramović (✉) · D. Šipuš
Faculty of Transport and Traffic Sciences, University of Zagreb, Zagreb, Croatia
e-mail: borna.abramovic@fpz.hr

© Springer Nature Switzerland AG 2020
M. Marinov and J. Piip (eds.), *Sustainable Rail Transport*, Lecture Notes in Mobility,
https://doi.org/10.1007/978-3-030-19519-9_2

main goal outlined is increasing the overall mobility of the population with emphasis on ecologically and economically efficient transport solutions.

Increasing the mobility of the population is not only the task for the urban, but also for peri-urban and rural areas. Usually, passenger railway transport is divided into international and national lines. National lines are then divided into long distance, regional, urban and suburban lines. Long distance lines connect country centres with each other and with regional centres. Urban and suburban lines aim to connect urban and suburban areas. The regional lines connect places inside the region. The focus of our research is on a regional line in the north-western part of the Republic of Croatia. For local populations, regional lines are very important because they provide everyday mobility option. Also, local populations expect adequate level of quality of service.

Quality of service is an assessment of provided service from the viewpoint of both service provider and the users. In this case study, the service provider is the railway passenger operator and the users are referred to as passengers. The feedback from passengers to the passenger railway operator is the crucial element for long-term planning and the provision of railway services. The feedback can be positive and negative, but both are important to improve the services. Regularly assessment of quality of service in a regional railway passenger context is a positive sign for the passenger, showing that operators are customer-oriented and that they are doing their best to keep up and improve the service.

2 Literature Review

A comprehensive approach to conducting surveys for the purposes of transport planning is presented in Richardson et al. (1995), 'Survey Methods for Transport Planning'. Tyrinopoulos and Aifadopoulou (2008) define quality of service with a wide range of characteristics in the transportation system services as provided to the passengers. These include safety, on-time performance, accessibility, efficiency, and many others. Their research provides an overview of a methodology developed by the Hellenic Institute of Transport to assess the levels of quality and performance of public transport services. Raised by Lalive et al. (2013) on whether increases in service frequency reduce road traffic externalities, they exploit the differences in service frequency growth by a procurement mode following a railway reform in Germany to address endogeneity of service growth. Their conclusive remarks identify that increases in service frequency reduce the number of severe road traffic accidents, carbon monoxide, nitrogen monoxide, nitrogen dioxide pollution, and infant mortality. Černa et al. (2017) proposed a methodology of selecting the transport mode for companies on the Slovak transport market where the main focus was freight transport. The criteria adopted in their methodology could be adapted for passenger transport research. The criteria comprised: (1) price for transport, (2) transport time, (3) transport safety, (4) reliability of carrier, (5) information, (6) flexibility, (7) additional services, (8) expertise and references, and (9) responsibility. Dolinayova

(2018) further researched possibilities of utilization of selected calculation methods for long-distance and regional railway passenger transport. Dolinayova et. al. (2019) also examines regional railway transport in the Slovak Republic using various indicators such as (1) transport performances, (2) operators, (3) regional railway line length, and (3) financing. This research shows interesting results that regional railway lines located in the regions with lower standards of living showed the best utilisation and the lowest loss for operators. Train passenger responses on provided services in case studies: PT Kereta Api (Indonesia) and Statens Järnvägar (SJ) AB (Sweden) were investigated by Saputra (2010). The research identified seven factors of service quality attributes that significantly influence customer satisfaction: (1) travel time, (2) information, (3) scheduling, (4) comfort, (5) tangible, (6) safety and security, and (7) service coverage. Dedik et al. (2017) made a comparative analysis of two connections between Bratislava and Sered in their research, which acknowledged point assessment for quality indicators. Gašparik et al. (2015) researched the quality in regional passenger rail transport with a comprehensive evaluation of the quality of the services in regional transport. Dell'Olio et al. (2011) conducted a study on how to use surveys to determine the desired level of service quality for users of public transport. Furthermore, Chang and Yeh (2002) conducted a research project using surveys in the field of air transport on the service quality of domestic flights. Marinov et. al. (2014) investigates the level of customer satisfaction as well as customer services in a passenger railway station Newcastle Central, UK. The results of this study indicate areas to investigate in the survey: (1) information, (2) infrastructure, (3) design, (4) accessibility, and (5) peak times. An important part of the survey is to keep the questions as simple as possible and open and closed questions are designed to be easily understood with limited options to answer. Some authors develop and employ a simple method for analyzing customer satisfaction surveys using different approaches and then employ a survey to collect the necessary data for analysis. Rüger (2019) researched the influence of passenger behaviour and passenger needs on the infrastructure facilities of stations. The results show that behaviour of passengers has a major influence on operational components such as (1) passenger flows in railway stations, (2) passenger exchange times and thus the (3) punctuality of trains. Moreover, Dolinayova (2011) investigated factors and determinants of modal split in passenger transport. In the field of railway passenger traffic, De Oña et al. (2014), used surveys, conducting research in northern Italy. The decision tree approach was used to analyse the data. Abramović et al. (2015) conducted a survey and comparative analysis in IC trains in Slovakia and Croatia. Abramović (2017) performed a passenger's satisfaction research on long distance terminals on the case study of the City of Zagreb. Stoilova (2018) conducted a study of railway passenger transport in the European Union. The study is a combination of a multicriteria model for rating railway passenger transport development. The factors for the classification include the following: (1) social and economic factors, (2) infrastructure factors, (3) factors associated with travel, and (4) technological factors. The learning from the literature review indicates: (1) levels of quality of services, (2) different criteria, and (3) survey techniques and analysis. This information is applied to this case study on regional railway line in the north-western part of the Republic of Croatia.

3 Case Study: Regional Railway Line Zabok–Đurmanec

3.1 Catchment Area

Krapina–Zagorje County is located in the north-western part of the Republic of
Croatia and represents a stand-alone geographical area. It is bordered by the mountain
peaks of Macelj and Ivančica in the north, the river Sutla in the west. The area is
separated from Slovenia, the basins of rivers Krapina and Lonja in the east and
the rest of the country. In the south-east the county is surrounded by the Zagreb
mountain Medvednica. The surface area of the county is 1229 km^2 with a population
of 132,892 (2011 Population census), and an average population density of 108
(Krapina–Zagorje County 2018).

The total track length in the county is 103,318 km, divided into regional and local
railway lines: R201 R201 Zaprešić–Čakovec (section Žeinci–Podrute) with a usable
length of 45.173 km; R106 Zabok–Krapina–Đurmanec–DG, 27.187 km; L102 Savski
Marof–Kumrovec–DG, (section Prosinec–DG) 16.783 km; L202 Hum Lug–Gornja
Stubica, 10.823 km and L201 Varaždin–Golubovec, (section Očura–Golubovec),
2.449 km. Along the Krapina–Đurmanec–DG track, the train speeds reach 50 km
per hour, with allowed axle load of 16 tons (Network statement 2018). The following official stops are located along the railway line: Zabok, Štrucljevo, Sveti Križ
Začretje, Dukovec, Velika Ves, Pristava Krapinska, Krapina, Doliće, Žutnica and
Đurmanec. The train stations are: Zabok, Krapina i Đurmanec. Krapina–Zagorje
County and railway lines are represent on Fig. 1.

The passenger carriage operates with Diesel railcars HŽ7121 and HŽ 7122. The
railcars are more than 35 years old, so at the end of lifespan, hence the purchase of
new Diesel railcars intended for regional passenger carriage, is essential in the near
future.

3.2 Survey Procedure

The participation in the survey was entirely voluntary. Therefore, the interviewers,
university students of railway transport, asked the respondents for permission before
starting the survey. The interviewers had undergone training that involved methods
and procedures of conducting a survey, rights of the respondents, and rights and
obligations of the interviewers. A special focus was put on the sensitive groups
of respondents (underage, mobility reduced, and senior passengers). A supervisor,
university professor of railway transport, was present during the survey.

The trains were open-spaced, when one passenger agreed to the survey, others
followed suit. The only passengers that did not participate were the ones sleeping or
exiting at the next stop/station.

Fig. 1 Krapina–Zagorje County and railway lines *Source* Adapted by the authors Krapina–Zagorje County (2018)

3.3 Questionnaire

The survey comprised a total of 27 questions, divided topically into two units: (1) general questions (5), and topical questions (22). There were 4 types of questions: 2 with free answer, 9 with single choice answer, 15 with numeric answers, and one multiple choice. We used a Google form sheet—the questionnaire was accessible via mobile phones (Android and iOS). The average duration of filling in the questionnaire was 4 min. Survey questions and question types are shown in Appendix.

3.4 Field Implementation of the Survey

The survey was conducted over the course of three days: Tuesday, 6 December, Wednesday, 7 December and Thursday, 8 December, 2016. There were a total of 13 participants and the survey was carried out in 20 passenger trains with the exception of early and late trains. There were 8 trains on the Zabok–Đurmanec Section (3124, 3126, 3128, 3130, 3132, 3134, 3136, and 3138), and additional 8 trains along the Đurmanec–Zabok Section (3117, 3119, 3125, 3127, 3129, 3131, 3103 and 3139), making it a total of 16 trains. A final number of 603 filled in questionnaires was obtained. Tuesday saw 248 participants, Wednesday 184, and Thursday 171.

88 B. Abramović and D. Šipuš

3.5 Analysis of Demographic and Social Economic Question

There were 339 male (56.2%) and 264 female (43.8%) respondents. The question on age structure was a single choice with a scale ranging from less than 18 years, between 18 and 24 years, between 25 and 34 years, between 35 and 44 years, between 45 and 54 years, between 55 and 64 years, and more than 65. 221 respondents (37%) were less than 18 years old, 146 respondents (24%) were aged between 18 and 24, 71 respondents (12%) between 25 and 34, 33 respondents (5%) between 35 and 44, 56 (9%) between 45 and 54, 41 respondents (7%) between 55 and 64 and 35 (6%) more than 65. Age structure of respondents is representing in Fig. 2. A significant decrease in passengers with age indicates that the majority, 61%, were pupils and students are using railway transport. The question on the highest level of education was a single choice with offered responses: *completed primary school, completed high school, completed college, a university degree, a Master degree of Science, and completed PhD.* 275 (46%) respondents had completed primary school, 211 (35%) respondents had completed high school, 55 (9%) respondents had finished college, 42 (7%) respondents had graduated from university, 17 (3%) respondents had obtained a Master of Science degree, and 3 respondents have obtained a doctoral degree. The structure of education coincides with the level of education at the national level with significant deviation towards higher educational attainment and higher levels of education. The completed education degree is in proportion to the level of income that actually represents passengers who use the public transport system for financial reasons.

A question on the monthly income was a single-choice question providing five options: (1) high (<€1000, (2) medium (€500–€1000), low (less than €500), (4) no income, and (5) I would rather not say. There were 20 (3%) responses for high, 106 medium, 96 low income participants (16%), 344 (57%) with no income and 37 (6%) of those who would rather not say. There is also a significant correlation between

Fig. 2 Age structure of respondents

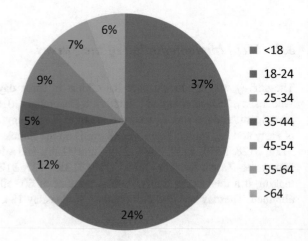

the income distribution and population age, as well as a direct correlation with the degree of education.

3.6 OD Matrix

Based on two questions about the origin and the destination of travel, we established an OD matrix for the passengers in the survey. A total of 603 passengers were interviewed in the trains on the line between Zabok and Đurmanec. Based on their answers obtained were 33 unique origins, 29 unique destinations, and 39 places of the start and end of the journey.

The survey in the trains forms a 39 times 39 OD matrix. Because of the complexity of presenting the matrix of 39 times 39, it is not presented in the paper. The matrix was filled in by searching the origins and filling in the matrix with destinations.

By analyzing the OD matrix in the trains, to and from Krapina, it may be concluded that in the implementation of the survey, 31.0% of passengers had Krapina as the origin of their journey whereas 33.2% of the passengers indicated Krapina as the destination of their journey. This result corresponds to the fact that Krapina is the centre of the Krapina Zagorje County.

Furthermore, interesting are the places that have a share greater than 2% as origins, and these are: Zabok (15.9%), Sveti Križ Začretje (11.6%), Đurmanec (7.0%), Zagreb (6.8%), Velika Ves (4.1%), Dukovec (3.5%), Doliće (3.3%), Žutnica (2.5%), and Donja Stubica (2.3%).

Also interesting are the places that have a share higher than 2% as destination, and these are: Zabok (23.2%), Zagreb (16.6%), Đurmanec (5.3%), Sveti Križ Začretje (4.5%), Bedekovčina (2.3%), and Velika Ves (2.3%).

3.7 Analysis of Questions About Quality of Service

The survey was conducted aboard the train and the origin/destination question refers to the final stop/station. An additional question was asked of respondents about how they arrived/departed from the stop/station, including the following options: bus/tram, personal vehicle, and on foot. Tram could have only been taken at Zagreb GK railway station and Zagreb ZK. Bicycles were used by 8 (1.3%) of the respondents, 56 took the bus (9.3%), 45 drove a car (7.5%), and 391 (64.8) walked. As for the departed from the final station/stop, there were 3 cyclists (0.5%), 57 took the bus (9.5%), 20 drove a car (3.3%) whereas 49 were passengers in a car (8.1%), while 474 (78.6%) walked. Considering the age structure, the walking part was expected, turning out at 2/3. It is also worth noting that the station is well-situated and accessible to pedestrians. A relatively high number of respondents (17.1%) arrive by cars with means it is too far for them to walk. Alternatively, they could have used the bus.

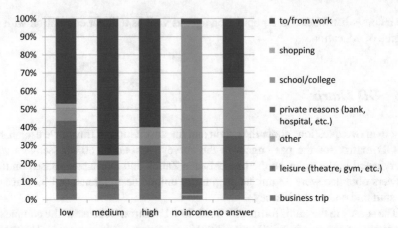

Fig. 3 Connection between average monthly income and purpose of the journey

The question "purpose of your journey" was a single-choice question for most of the respondents, 311 (51.6%), the purpose was traveling to/from school/college, for 161 (26.7%) to/from work, 83 (13.8%) travelled for private reasons (bank, hospital, etc.), 18 (3.0%) were on a business trip, 16 (2.7%) leisure (theatre, gym, etc.), and for other travel purposes there were 8 (1.3%), and for shopping only 6 (1.0%) respondents. 51.6%, or more than half, of the respondents cited going to/returning from school or college, while only 13.8% stated arriving to/going to work. We identified that, public passenger transport in Croatia, mainly deals with the transport of pupils and students who have largely subsidised transport. There has, for last few years, been a state and county-funded subsidy, so pupils and students travel at no cost. For future development of the public transport system this is a significant cost and a great challenge to recoup funds in a different manner. Most journeys are the ones including commuting to school/college–as many as 47.9% of the respondents, while the groups low, medium and high there is a high number of commuters to/from work, 7.5, 13.3 and 2%, which amounts to 22.8%. It is necessary to point out that purpose of journey school/college includes 92.9% with no income and 3.2% passengers that did not give an answer. But rest of respondents claim that they have income. Analyzing this answer according to the age, identifies that passengers are going to additional activities at school/college. Figure 3 shows the connection between the average monthly income and the purpose of the journey.

Questions on the reasons why respondents use rail as a mode of transport was a multiple choice type with following options: (1) accessibility, (2) financial reasons, (3) another reason, (4) regularity, (5) comfort, (6) speed, and (7) punctuality. The majority of passengers have chosen accessibility and financial reasons as to why they are using rail. As pointed out earlier, railway stations are well-situated and very convenient for passengers to opt for railway transport, making accessibility a reasonable factor. Keep in mind that majority of passengers are pupils/students and

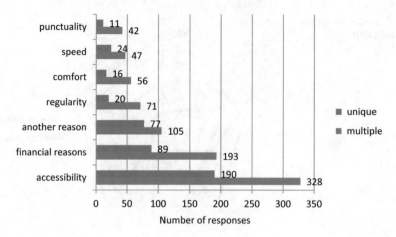

Fig. 4 Reasons why you use the railway as a means of transport

they do not pay direct fares and this is a strong reason to use rail. Detailed results are shown in Fig. 4.

Frequency of travelling by train has been examined with the question on 'how often one travels by train in one direction per week'. Ten times travelling per week correspondents with five trips per week. Overall, 48.6% of passengers take 5 trips per week and 8.6% of passengers take 6 trips per week. Interestingly, 12.6% of passengers take only one trip per week and 6.6% have made three trips per week. The number of trips correlates with the purpose of travel so 66.9% purpose of the trip "school/college" have 5 trips per week. It is an interesting purpose of the journey "to/from work" that 43.5% of working people have 5 journeys per week, and 24.8% have 6 trips per week. The questionnaire had eleven numeric questions to evaluate passenger satisfaction. These questions were a single-choice type with the scale ranging from 1 (*poor*) to 5 (*excellent*).

Travel comfort was rated poorly by 33 (5.8%) respondents, satisfactory by 86 (14.3%), good by 213 (35.3%), very good by 193 (32%), and excellent by 78 (12.9%). The overall travel comfort was rated as good for more than a third of the passengers. One of the reasons is the more than 35 years old railway diesel multiple units are at the end of their lifespan. That means that on one hand it's very uncomfortable for passengers and on other hand very challenging for maintenance work.

Time of the trip was rated poorly by 57 (9.5%) respondents, satisfactory by 94 (15.6%), good by 164 (27.2%), very good by 159 (26.4%), and excellent by 129 (21.4%). Distribution of time travel is almost uniform across categories. More than 50% of passengers evaluate trip time as good and very good.

The possibility of having a seat in the train was rated poorly by 24 (4.0%) respondents, satisfactory by 72 (11.9%), good by 126 (20.9%), very good by 199 (33.0%), and excellent by 182 (30.2%). Exactly one third of passengers evaluated available seats as very good, and more than 30% evaluated available seats as excellent. We

Fig. 5 Grades for station
and train staff

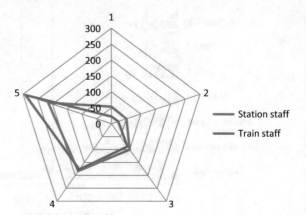

can draw a conclusion that the capacity of trains is adequate or, in different point of
view, that demand for seating places is satisfied.

Trip price was rated poorly by 26 (4.3%) respondents, satisfactory by 48 (8.0%),
good by 142 (23.5%), very good by 175 (29.0%), and excellent by 212 (35.2%). The
result is obvious because of age structure of passengers who are not students, and
therefore not subsidised, and purpose of journey.

The process of buying the ticket was rated poorly by 25 (4.1%) respondents,
satisfactory by 69 (11.4%), good by 124 (20.6%), very good by 182 (30.2%), and
excellent by 203 (33.7%). The majority of passengers have a monthly pass which
explains the high grades for process of buying a ticket. They are using ticket office
only once per month.

Station staff was rated poorly by 54 (9.0%) respondents, satisfactory by 51 (8.5%),
good by 100 (16.6%), very good by 183 (30.3%), and excellent by 215 (35.7%). Train
staff was rated poorly by 23 (3.8%) respondents, satisfactory by 23 (3.8%), good by
88 (14.6%), very good by 177 (29.4%) and excellent by 292 (48.4%). Station and train
staff are railway workers who communicate regularly with passengers and represent
the public appearance of the railway company. The analysis of rating is excellent
for the railway company because more than one third of passengers rate station staff
and almost half of passengers rate train staff as excellent. Figure 5 show the grades
between station and train staff.

Punctuality was rated poorly by 35 (5.8%) respondents, satisfactory by 57 (9.5%),
good by 128 (21.2%), very good by 208 (34.5%), and excellent by 175 (29.0%). More
than one third of the passengers have rated punctuality as very good. This is a very
high rate for the railway service in Croatia, as trains operate only on this line, so
there is no interface with other trains on the network and delays are likely only when
something unpredictable happens.

The frequency of departure was rated poorly by 40 (6.6%), satisfactory by 79
(13.1%), good by 180 (29.9%), very good by 179 (29.7%), and excellent by 125
(20.7%). Interestingly, this is almost the same number of passengers who rate fre-

Table 1 Grade for average and mode for all questions about quality of service

Question	Short description	Average	Mode
1	Travel comfort	3.3	3
2	Trip time	3.3	3
3	Available seats	3.7	4
4	Trip price	3.8	5
5	Ticket buying	3.8	5
6	Station staff	3.8	5
7	Train staff	4.1	5
8	Punctuality	3.7	4
9	Frequency of departure	3.4	3
10	Fit of timetable	3.5	5
11	Train information system	2.8	1

quency as good and very good. Almost one quarter of passengers are unsatisfied with the frequency of trains. This data reveals that there is place for improvement of the frequency of train departures.

The question "How does the timetable line fit your needs?" was rated poorly by 61 (10.1%), satisfactory by 68 (11.3%), good by 151 (25.0%), very good by 146 (24.2%), and excellent by 177 (29.4%) respondents. More than one quarter of passengers are dissatisfied with the timetable. In fact, the timetable has remained unchanged for the last 30 years. Surely, in such a large period, there were huge changes in demand.

The train information system was rated as poorly by 165 (27.4%), satisfactory by 94 (15.6%), good by 129 (21.4%), very good by 124 (20.6%), and excellent by 91 (15.1%). The only information system is physical, in the form of train staff. They have a very limited source of information for unplanned and unpredicted situations and have to rely on their own knowledge. Indeed, there is a necessity for implementation of a new on-board train information system.

Keeping in mind all eleven questions about the quality of service, the average grade is 3.6 and the mode grade is 4. The total grade is, therefore, 4 or very good. Table 1 lists average and mode grade for all questions about service quality. The worst grade is given to Train information system. The main reason is that there is only one ticket inspector aboard the train. They give information from time to time but, according to the passengers, this is completely inadequate. An illustrative example that a passenger put forward during the survey was that the train was delayed for more than 10 min, and the information from the train staff was that the train was delayed for 5 min. Using today's technology, the train information system is a very simple and low-cost solution.

As many as 85% of the passengers believe that Wi-Fi services are needed aboard the trains. This is very reasonable and completely in line with technological advances in other countries. Only 58% of passengers believe that the railway is an environmentally conscious mode of transport. Although the media, politicians and scientists talk about the ecological benefit of the railway, the passengers do not recognize it. It is interesting to note that in the conversation with passengers they pointed out that train runs on diesel, and in their opinion, that is not environmentally friendly.

The question "Will you use integrated ticket?" was answered positively by 85% of the respondents. This fact undoubtedly confirms, that is a greater need for the introduction of an integrated passenger transport system and a uniform transport ticket.

4 Conclusion

Population mobility has been a hot topic in transport for the last few years. Increasing the mobility is not only the task for urban, but also for peri-urban and rural areas. The focus of our research was on regional lines in north-western part of the Republic of Croatia. For the local population, the regional lines are very important because they provide everyday mobility options. Local populations expect an adequate level of service quality because they pay for the service. Regular assessment of service quality in railway regional passenger transport has revealed a positive sign for passengers because the operator is customer-oriented and doing his best to keep up and improve the services.

Our research was conducted on the railway regional line in the northwest part of Croatia - the railway line between Zabok and Đurmanec. This line has every characteristic of a passenger regional railway line in central part of Europe. Our findings suggest that there is need for regular quality assessment. For sure passengers in this area need rail transport; it is a public service obligation (PSO).

Reasons why the survey respondents use rail as a mode of transport was a multiple-choice question with following options: (1) accessibility, (2) financial reasons, (3) another reason, (4) regularity, (5) comfort, (6) speed, and (7) punctuality. The majority of passengers have chosen accessibility and financial reasons as to why they choose rail transport.

Regarding all the eleven questions about the quality of service, the average grade was 3.6 and the mode grade was 4 (very good), which makes the total grade very good. The worst grade was attributed to the train information system. The main reason behind that, is because there is only one ticket inspector aboard the train who provides information from time to time and according to the passengers, this is completely inadequate. In today's technology, train information systems are a simple and low-cost solution. As many as 85% of the passengers believe Wi-Fi services are needed aboard. This is very reasonable and completely in line with technological advancements in other countries. Only 58% of the passengers believe that the railway is an environmentally conscious mode of transport with 85% of respondents confirming they would like to have an option for buying an integrated transport ticket for different modes of transport and different operators. The paper confirmed there is a great need for the introduction of an integrated passenger transport system.

Appendix: Survey Questions

	Questions	Questions types
1	Date	Numeric
2	Train number	Numeric
3	Sex	Single choice
4	Age	Numeric
5	Highest level of education	Single choice
6	Average monthly income	Single choice
7	Origin	Free text input
8	How you have reach your origin	Single choice
9	Destination	Free text input
10	How you proceed from your destination?	Single choice
11	Purpose of the journey	Single choice
12	Specify the reasons why you use the rail as a means of transport?	Multiple choice
13	How often do you travel on trains?	Numeric
14	Travel comfort	Numeric
15	Trip time	Numeric
16	How do you rate the number of available seating areas?	Numeric
17	Trip price	Numeric
18	Way of buying the ticket	Numeric
19	Station staff	Numeric
20	Train staff	Numeric
21	Punctuality	Numeric
22	Frequency of departure	Numeric
23	How does the timetable line fit your needs?	Numeric
24	Train information system	Numeric
25	Wi-Fi	Single choice
26	Ecology	Single choice
27	Will you use integrated ticket	Single choice

References

Abramović B (2017) Passenger's satisfaction on long distance terminals: case study city of Zagreb. Period Polytech Transp Eng 45(1):42–47. https://doi.org/10.3311/PPtr.9197

Abramović B, Nedeliakova E, Lukinić D (2015) Comparative analysis on providing the service in ic trains between Republic of Slovakia and Republic of Croatia. Int J Traffic Transp Eng 5(3):309–318. https://doi.org/10.7708/ijtte.2015.5(3).08

Černá L, Zitrický V, Daniš J (2017) The methodology of selecting the transport mode for companies on the Slovak transport market. Open Eng 7(1):6–13

Chang YH, Yeh CH (2002) A survey analysis of service quality for domestic airlines. Eur J Oper Res 139(1):166–177. https://doi.org/10.1016/S0377-2217(01)00148-5

De Oña R, Eboli L, Mazzulla G (2014) Key factors affecting rail service quality in the Northern Italy: a decision tree approach. Transport 29(1):75–83. https://doi.org/10.3846/16484142.2014. 898216

Dedík M, Gašparík J, Záhumenská Z (2017) Quality assessment in the logistics of rail passenger transport. In: MATEC web of conferences. EDP sciences, vol 134, p 00009. https://doi.org/10. 1051/matecconf/201713400009

Dell'Olio L, Ibeas A, Ceca P (2011) The quality of service desired by public transport users. Transp Policy 18(1):217–227. https://doi.org/10.1016/j.tranpol.2010.08.005

Dolinayova A (2011) Factors and determinants of modal split in passenger transport. Horiz Railw Transp Sci Pap 2(1):33–39

Dolinayova A (2018) Possibilities to quantify fixed costs for long-distance and regional railway passenger transport. In: Proceedings of ICTTE 2018. University of Belgrade

Dolinayova A, Danis J, Cerna L (2019) Regional railways transport—effectiveness of the regional railway line. In: Fraszczyk A, Marinov M (eds) Sustainable rail transport. Lecture notes in mobility. Springer. https://doi.org/10.1007/978-3-319-78544-8_10

European Commission (2011) Roadmap to a single european transport area: towards a competitive and resource efficient transport system: White Paper. Publications Office of the European Union

Gašparík J, Stopka O, Pečený L (2015) Quality evaluation in regional passenger rail transport. Naše More 62(3):114–118. https://doi.org/10.17818/NM/2015/SI5

Krapina–Zagorje County (Krapinsko zagorska županija) (2018) Available from the Internet: www. kzz.hr

Lalive R, Luechinger S, Schmutzler A (2013) Does supporting passenger railways reduce road traffic externalities?. CEPR discussion paper no. DP9335. Available at SSRN: https://ssrn.com/abstract=2215454

Marinov M, Lima T, Kuhl B, Bogacki A, Onbasi C (2014) Analysis of customer services in railway passenger stations using a holistic method-application to Newcastle Central station. Transport Problems, 9(S):61–71

Network statement (2018) HŽ Infrastrucutre Ltd. Available from the Internet: http://www.eng. hzinfra.hr/wp-content/uploads/2018/06/2018-Network-Statement-I.-II.-and-III.-modification. pdf

Richardson AJ, Ampt ES, Meyburg AH (1995) Survey methods for transport planning. Eucalyptus Press and University of Melbourne, Parkville, p 475

Rüger B (2019) Influence of passenger behaviour on railway-station infrastructure. In: Fraszczyk A, Marinov M (eds) Sustainable rail transport. Lecture Notes in Mobility. Springer https://doi. org/10.1007/978-3-319-78544-8_8

Saputra AD (2010) Analysis of train passenger responses on provided services (Case study PT. Kereta Api Indonesia and Statens Jarnvargar (SJ) AB Sweden). Master's thesis. Karlstads Universitet. 127 p

Stoilova S (2018) Study of railway passenger transport in the European Union. Tehnički vjesnik 25(2):587–595. https://doi.org/10.17559/TV-20160926152630

Tyrinopoulos Y, Aifadopoulou G (2008) A complete methodology for the quality control of passenger services in the public transport business. Eur Transp/Trasp Eur 38(2008):1–16

The Possibilities of Increasing the Economic Efficiency of Regional Rail Passenger Transport—A Case Study in Slovakia

Anna Dolinayova and Lenka Cerna

Abstract Short-distance rail passenger transport is regarded as a service of public interest in almost all EU countries. With regard to European Transport Policy and measurements proposed in the Fourth Railway Package, it is necessary to improve the quality of rail passenger transport services, so it reflects customers' requirements, and the optimal utilisation of public resources. Opening the market of this rail passenger transport segment may, however, not bring expected results. The efforts to reduce a drain on public resources may induce a lower quality of services and lower performance of rail services. Public passenger transport represents only one quarter of the total share of passenger transport on the Slovak transport market where the share of rail passenger transport is approximately 9.4%. A separate issue is the impact of a zero fare, introduced in November 2014, on the attitude of the public on rail transport. As a result, the interest of the public in travelling by passenger trains has increased, but at the same time, the quality of provided services has been impacted significantly. This paper outlines how selected technical, technological and other factors influence the efficiency of providing services within a rail passenger transport segment in a selected urban agglomeration. The results suggest; applying a tariff system and setting quality standards of services for rail passenger transport.

Keywords Rail transport · Short-distance · Passengers · Efficiency · Tariff system · Quality

1 Introduction

The analysis of rail transport development in the EU and the Slovak Republic, in the context of strategic transport-political objectives, points to positive development in rail passenger transport. The declaration of a state transport policy in the Slovak Republic is similar to the principles of the EU. The emphasis is on utilising rail

A. Dolinayova (✉) · L. Cerna
University of Zilina, Univerzitna 8215/1, 010 26 Zilina, Slovakia
e-mail: anna.dolinayova@fpedas.uniza.sk

© Springer Nature Switzerland AG 2020   97
M. Marinov and J. Piip (eds.), *Sustainable Rail Transport*, Lecture Notes in Mobility,
https://doi.org/10.1007/978-3-030-19519-9_3

transport as a more environmentally friendly mode of transport (Strategy PPT 2020 2015).

The measurements of the Fourth Railway Package contribute to creating a more efficiently working industry which better responds to customer needs and improves the relative attractiveness of the rail transport industry concerning other modes of transport (Fourth railway package 2016). All measurements in the Fourth Railway Package make it easier for new rail operators to enter the rail transport market. Synergies will be achieved through a combined effect of individual initiatives. The efficiency of opening the market is dependent on the introduction and efficient application of certain "frame conditions", such as non-discriminatory access to infrastructure, access to appropriate rail locomotives, access to stations or assignment of a train route including the traffic management. Some of these frame conditions will be solved through initiatives related to the market opening with services of domestic rail passenger transport. The other ones will be solved through a proposal related to strengthening and empowering infrastructure management. To reduce obstacles for new participants entering the market, some synergies should be achieved by setting simplified procedures for issuing safety certificates for rail enterprises as well as permissions to put rail vehicles on the market (Strategy for DTI by 2020 and Fourth Railway Package 2016).

Liberalisation of the passenger transport market in the Slovak Republic has not resulted in a higher share of rail transport in the potential transport market. The increase of passengers on trains in the Slovak Republic has been evoked mainly due to subsidised fares as a result of Government regulations (free transport) as well as competitors entering the market with low prices. The increase of a rail transport share is based on a combination of a competition and commercial principles, orders for public services, and regulatory measurements (Strategy for DTI by 2020).

The methodology used in this paper is divided into several parts, and its primary goal is to study the possibilities of increasing economic efficiency of regional rail passenger transport through the optimal establishment of processes on the analysed part of the railway line in question. This will bring cost savings and ultimately increase economic efficiency while preserving quality standards of public passenger rail transport. In order to achieve this scientific method, individual steps will be used. The first part of the methodology describes the observed problem, which deals with the analysis of the development of public passenger railway transport in the Slovak Republic. We will examine selected factors influencing the efficiency of services provision in the railway passenger transport segment in the selected agglomeration. During the research, the number of passengers transported in individual modes of transport is analysed.

Passengers traveling free are also studied. Collected information, complexity and depth of examined indicators were the basis for creating a research plan which identified individual phases, methods and procedures. The following part of the methodology contains an analysis of the occupancy of the trains and the analysis of the timetable on the selected part of the railway line. On the basis of these analyses, the possibility of optimising the train journeys and the circuit of the rail vehicles need for the service operation on the selected part of the railway line is identified. The conclusion of the methodology is supplemented by the calculation of the economic efficiency of rationalisation measures based on the monitoring of the cost of the proposed solution. The methodology represents a sequence of steps to ensure an increase in the economic efficiency of rail passenger transport in the monitored region.

2 Literature Review

Many authors have been dealing with the issue of economic efficiency of public passenger transport and associated influencing factors. Fitzová et al. (2018) identified factors impacting the efficiency of urban public transport systems in the Czech Republic. An innovative approach to the evaluation of public bus transport efficiency was introduced by Avenali et al. (2018) who tried to apply a hybrid model of costs in order to determine standard unit costs of local public bus transport. As the authors state, the results of such an evaluation may be a useful foundation for defining financial compensation of services within local public bus transport.

The economic efficiency of a railway system has recently been a topic discussed by economists in every European country as well as in countries out of Europe where rail transport works well (Catalano et al. 2019; Dolinayova et al. 2017). The productivity of the railway systems is constrained by the existing rail infrastructure (Marinov et al. 2013). Stoilova (2018) researched the development of rail passenger transport in the European Union countries by using criteria related to the transportation process and the level of economic development of the countries. In her study, she proposed a methodology based on a combination of multi-criteria methods with economic and social factors selected as primary criteria. Fraszczyk et al. (2016) studied various indicators impacting railway performance using available statistic data. This study identified the economic efficiency of regional railway transport could be increased by more effective utilisation. Simulation models can be used to analyse the use and productivity of railway lines (Singhania and Marinov 2016). On the other hand, the most critical factors in the use of railway transport relate to passenger behaviour (Rüger, 2017), travel time (Weerawat et al. 2017) and other quality factors.

The Fourth Railway Package (Fourth Railway Package 2016) proposes many measurements for increasing railway performance. Individual countries have been introducing some reforming measurements which contribute to stronger competitiveness of a rail transport system including a shift of modal split in favour of the railways. However, have the reforms brought an expected effect? Studies of single authors do not show precise results. Smith et al. (2018) studied the impact of reforms in rail transport on its efficiency with results suggesting that the cost reduction arising from a regulatory reform depends on the degree of the market opening (actual or desired), vertical structure and intensity of the network usage.

Reforms in individual EU countries may not always bring positive effects in the area of costs. Growitsch and Wetzel (2009) analysed 54 railway companies from 27 European countries. Their research showed that "integrated railway companies are, on average, relatively more efficient than "virtually" integrated companies and that a majority (65 percent) of the railway companies observed indicate economies of scope" (Growitsch and Wetzel 2009). Bogart and Chaudhary (2015) studied the influence of state ownership on railway performance in India. Their results showed that state ownership was not worse than private ownership. Despite the majority of documents showing that introducing the competitiveness in rail transport (on a passenger and freight market) has had a positive impact from efficiency and productivity points of view, the study of the impact of vertical separation has brought various results (Cantos et al. 2012). Implementation of reforms may, however, increase the attractiveness of rail transport to customers and bring synergistic effects of the liberalisation of the railway transport market (Panak et al. 2017).

Considering that the economic efficiency of passenger rail transport is different in individual countries, the implementation of reforming measurements does not always increase efficiency. Therefore, we have decided to research the options of its increase, no matter if services are provided by the state or a private carrier. We performed an analysis of public passenger transport in the Slovak Republic using a case study in a selected agglomeration. We demonstrated how the costs of providing short-distance rail passenger services could be reduced based on non-investment measurements.

3   Analysis of Public Passenger Rail Transport in the Slovak Republic

A basic prerequisite for increasing the attractiveness of public passenger transport is the need to improve individual connections and links, by individual carriers, in all transport modes in the whole territory. Public passenger rail transport in Slovakia is a result of European Transport Policy, Principles of State Transport Policy of the Slovak Republic and the Transport Development Strategy of the Slovak Republic by 2020. It conforms to Regulation (EC) No 1370/2007, Act No 513/2009 Coll for the railways and amendments of some acts and Act No 514/2009 Coll on transport on the railways (Strategy for PPT by 2020 2015).

The Transport Development Strategy of the Slovak Republic by 2020 (a resolution of the Government of the Slovak Republic No 158 of 03 March 2010) is an underlying document which defines fundamental long-term goals, priorities of transport development in the Slovak Republic, and tools and resources required to achieve the goals. It represents a ground for preparing other conceptual ideas and vision of the Ministry of Transport and Construction of the Slovak Republic for formulating the position of the Slovak Republic towards a future European Transport Policy. The Transport Strategy complies with conceptual materials adopted at EU levels, such as the Lisbon Strategy, the Gothenburg Strategy and the Transport Policy of the EU.

Private carriers organise rail passenger transport in the territory of the Slovak Republic in cooperation with the Ministry of Transport and Construction of the Slovak Republic. In some relatively small areas of self-governing regions, central ordering of rail transport enables efficient operating of regional transport. Regional transport is currently provided by the following carriers: Železničná spoločnosť Slovensko ("ZSSK, a. s.") and RegioJet ("RJ"). The planning of regional transport is based on Economic and Social Development Plans, Territorial Plans, and (if worked out) Transport Operation Plans and Territorial Generals of Transport (Daniš et al. 2016).

3.1 Support of the European Transport Policy for the Development of Public Passenger Rail Transport

For more than 25 years transport policy of the EU has been oriented at building a sustainable and competitive transport system. Transport-political measurements, to support rail transport in order to increase its share on the transport market, have been integral parts of the EU policies for the long term. Rail transport has ambitions to significantly contribute to the fulfilment of the EU strategy which is mainly focused on increasing the competitiveness as well as durable economic, social, environmental and energetic sustainability, cohesion and safety (White Paper 2011).

In the White Paper 2011 on Transport the Commission, the EU introduced its vision for the provision of a Single European Rail Area (SERA), establishing an approach to ensure transport competitiveness in the EU in a long-term term. At the same time, the Commission dealt with issues of expected growth, fuel safety and CO^2 emissions reduction. An essential aspect of this policy is to strengthen rail transport role regarding difficulties of reducing the dependency on oil in other industries. This strategy can, however, be achieved only if rail transport provides efficient and attractive services. If failures in regulation and the market are eliminated, including obstacles to enter the market, and complicated administration procedures, efficiency and competitiveness of rail transport will proceed.

Building a single rail area represents one of the options to make use of rail transport potential. European Transport Policy for rail as a sustainable transport option, depends on wide-ranging tools. Regulation of the transport market and creation of conditions to favour rail and water transport are based on effects of such financial and non-financial tools which support transport liberalisation, do not infringe the competition and ensure the interoperability.

The EU transport policy aims to build an inner market with rail transport and thus to create a more efficiently working industry with better reactions to customer needs. Therefore, earlier EU legal regulations document some underlying principles. These principles regulate the improvement of rail transport efficiency through a progressive market opening, independent rail enterprises establishment, infrastructure managers and their accounts separation. The principles have progressively been applied since 2000. Lastly, adopting three subsequent packages of the EU legal regulations have made some impact, but the total share of rail transport in the EU within utilised transport modes remains low. Partially, it is due to issues with appropriateness (e.g. rail transport is not convenient from the point of view of many journeys for short distances within a town, such as shopping in supermarkets), but also as a result of obstacles to enter the market which restrict competition and innovations (White Paper 2011).

Concurrently, with opening the rail transport market, other EU measurements have improved the interoperability and safety of domestic networks. A more European approach to rail transport should simplify cross-border movement so rail transport can utilise its competitive advantage for longer distances and ensure lower costs for a single market with suppliers of rail equipment.

The Fourth Railway Package was designed to complete a single European rail area and to improve interoperability. It aimed to enable the competition related to agreements on public services on domestic markets in order to improve the quality and efficiency of services in the area of domestic passenger transport.

There are the following specific measurements of the Fourth Railway Package (Fourth Railway Package 2016):

• infrastructure management,
• opening the market with services of domestic passenger rail transport,
• interoperability and safety, and
• social scale.

The Fourth Railway Package introduces an integrated approach which will create conditions for growth of the entire rail transport system. Through progressively increasing reliability and efficiency, it will also increase its share on the market, set up conditions for better quality services and enable rail transport to develop its unused potential, so it becomes a real and attractive alternative. The Fourth Package concentrated on finishing the single rail area and was adopted in April 2016 (a technical pillar) and in December 2016 (a market pillar) (Fourth Railway Package 2016).

3.2 Historical Development of Public Passenger Rail Transport in the Slovak Republic

Rail transport was the first means to provide mass public passenger transport in the territory of the Slovak Republic (Strategy for PPT by 2020 2015). In the inter-war period (years) rail carriers also provided services of public passenger transport by road (Strategy for PPT by 2020 2015). Since 1949, the company, Czechoslovak State Railways provided only rail transport services. Until the dissolution of the Czech and Slovak Federative Republic on 01 January 1993, Czechoslovak State Railways were controlled by the Transport Department (Strategy for PPT by 2020 2015).

After dissolving the company Czechoslovak State Railways into two entities, a state-owned company Železnice Slovenskej Republiky ("ŽSR") was formed in Slovakia. On 30 September 1993, the Act No 258/1993 on ŽSR was enacted. The adopted act defined ŽSR in Slovakia as a state-owned company applying elements of commercial and public service management, the only one of its kind. In connection with the economic recovery of the railways in Slovakia and integration of the ambitions of Slovakia into the EU, the resolution of the Government of the Slovak Republic No 830 approved the "Transformation and Restructuring Project of ŽSR" in 2000. The Project proposed a physical and accounting separation of the management and operation of a railway from transport and business activities. On its grounds, the assets and activities of the carrier were separated from a railway operation, and on 1 January 2002, two railway companies were created—Železničná spoločnosť, a. s., as a provider of transport and business activities, and Železnice Slovenskej Republiky as an infrastructure manager. The founder and 100% shareholder of the new company Železničná spoločnosť, a. s., was the Slovak Republic, where the Ministry of Transport, Post and Telecommunications in the Slovak Republic acted on its behalf. At the turn of 2004 and 2005, in compliance with the resolution of the Government of the Slovak Republic No 662/2004, that company Železničná spoločnosť, a. s., was transformed into two independent companies, and at the same time, passenger and freight rail transport were separated. The role of a provider of passenger rail transport services was taken over by the newly created company Železničná spoločnosť Slovensko, a. s., ("ZSSK"), and the role of a provider of freight rail transport services was taken over by the newly created company Železničná spoločnosť CARGO Slovakia, a. s (Strategy for PPT by 2020 2015).

Thus, three independent railway companies have been acting in Slovakia since 1 January 2005: Železnice Slovenskej Republiky (ŽSR), Železničná spoločnosť Slovensko, a. s. (ZSSK), and Železničná spoločnosť Cargo Slovakia, a. s. (ZSSK Cargo), which emerged as a result of a progressive transformation of a joint-ventured enterprise. Since 4 March 2012, the operation of passenger rail transport on the Bratislava–Komárno track has been taken over by the private carrier RegioJet a. s. which replaced the company Železničná spoločnosť Slovensko, a. s. The market opening in the Slovak Republic caused the joining of new rail carriers as providers of transport services in public passenger transport—in 2014 Leo Express and in 2016 ARRIVA entered the market.

3.3 Operation of Public Passenger Rail Transport in the Slovak Republic

Rail passenger transport in the Slovak Republic is mainly provided by the state-owned company Železničná spoločnosť Slovensko, a. s., (transport performance of 94.4% (Statistical Office of SR 2018) and other private rail carriers [transport performance of 5.6% (Statistical Office of SR 2018)]. The transport performance is expressed using the number of carried passengers in the observed year of 2016. Železničná spoločnosť Slovensko, a. s., transported 65,606 thousand passengers in 2016 and the rest of carriers operating in rail passenger transport transported 3,919 thousand passengers. As a result of free transport, implemented in 2014, a significant growth of carried passengers was noticed in the following years (2015–2016). In 2014, public passenger rail transport was used by 49,272 thousand passengers, in 2015 it was 60,566 thousand passengers, and in 2016 it was 69,525 thousand passengers (Statistical Office of SR, 2018).

Suburban rail transport is characterised with short transport distances and a high number of passengers. It provides daily transport of people from their legal residence or temporary address to work, school or somewhere else for personal purposes (doctor and/or office appointment, cultural events) from a subregion into big towns (Masek et al. 2015). ZSSK demonstrated that 30% of the population travel every workday and 17% travel several times a week which forms the greatest share in suburban transport (70%). This sector of passengers is susceptible to frequent changes in train schedules.

Regional transport is considered as transport among subregions of towns within a natural historical or administratively bounded region. Under conditions of the Slovak Republic, this transport is framed by regional boundaries, and according to ZSSK researchers, forms approximately 30% of all journeys. The segment responsible for passenger trains "Os" (the lowest category of passenger trains as mostly they provide transportation for short distances (regional services) and stop at all stations and stops) are in broader surroundings of the most important towns (Bratislava, Košice, Žilina, Prešov) with the most significant area of subregion. They also operate express trains, or trains of REX category (category of trains of the carrier Železničná spoločnosť Slovensko, a.s. and they usually have the same stops and go in line with an express train, but have a shortened route over them). The main groups of customers are those who commute daily to work or school, who attend health care institutions, offices or travel in their free time. Seasonal fluctuations account for transport of tourists and passengers travelling in their free time, reaching up to 20% (Strategy for PPT by 2020 2015). An increase in the number of passengers occurs mainly in tourist regions (a special position has the High Tatras Region of tourism) in the summer period (Strategy for PPT by 2020 2015).

Long-distance domestic transport has the function of transport to distant towns. In this segment, a significant share of travellers are students (20%) and weekly commuters travelling for a full fare, forming up to 25% out of the total number of passengers (Strategy for PPT by 2020 2015).

International long-distance transport includes transport of people to and from the territory of the Slovak Republic, and transit of people through the territory of the Slovak Republic to neighbouring or distant countries, and back. After the Slovak Republic's entry into the EU, international transport has been growing. International transport is provided with trains of IC (type of express train that operates in domestic traffic), EC (international trains higher category), EN (international night trains higher category), Ex (international or domestic trains higher category) and R categories (domestic day and night long-distance trains) in compliance with agreements with the relevant railways. International transport focuses on connecting regional centres and the capital of the Slovak Republic with capitals and industrial and urban agglomerations in neighbouring countries. During discussions with foreign railway management, efforts are on negotiating international trains which have reasonable occupancy in the territory of the Slovak Republic (Strategy for PPT by 2020 2015).

3.4   Regional Rail Transport in the Slovak Republic

On the market of domestic passenger transport in the Slovak Republic, there are currently two carriers: Železničná spoločnosť Slovensko, a. s., (ZSSK) and RegioJet a. s. (RJ). The state entered into agreements on transport services in the public interest with these two companies (Public service transport contracts in the operation of passenger transport on the track, 2018). The Ministry of Transport and Construction of the Slovak Republic orders transport performance in regional and long-distance rail passenger transport. Currently, the government allows a carrier to provide transport services at their own business risk on the tracks of Železnice Slovenskej Republiky (ŽSR). This principle is applied on rail links with sufficient volumes.

Funding for rail passenger transport is realised on the basis of the "Agreement on Transport Services in Public Interest" (pursuant to Act No 514/2009 Coll on the transport on the railways, pursuant to Act No 164/1996 Coll on the railways—Agreement on Performance in Public Interest until 31 December 2009), under which the state represented by the Ministry of Transport and Construction of the Slovak Republic, reimburses a provable loss from public rail passenger transport in the territory of the Slovak Republic. The agreement is binding for the state in terms of the obligation to reimburse the loss to the carrier. Such a loss in passenger transport means a difference between economically justifiable costs spent by the carrier in order to fulfil the commitment under the agreement including adequate profit and incomings gained through this commitment by the carrier.

As a result of free transport, more and more Slovaks have been travelling by train (Majerčák and Černá 2015). The state-owned carrier Železničná spoločnosť Slovensko (ZSSK) records an increasing number of passengers carried (Majerčák and Černá 2015). While many people utilise trains mainly for longer distances, transport for short distances and regional tracks not currently utilised, are also attractive. The state has partially reopened the operation on three of them in recent months. The Transport department, however, claims that it has finished with this initiative number for the time being (Majerčák and Černá 2015).

The possible reopening of passenger rail transport on tracks where transport is currently discontinued is contingent on having a sufficient flow of passengers. Such tracks will allow utilisation of the railways to their full extent during the entire day (Gasparik et al. 2018). Passenger rail transport operations are efficient in case of transporting significant passenger flows. In other cases, mobility gained by utilising bus transport.

Prices on rail passenger transport are regulated on the basis law by the Transport Authority. The exception is the price for commercial transport services; under the conditions of the Slovak Republic and to the current date we speak about IC trains of ZSSK, a. s. Until 2012, IC trains (definition in Sect. 3.3) were operated under the Agreement on Transport Services in Public Interest. Since 2012, IC trains of the state carrier have been of commercial interest.

Basic fares and special fares for transport services are set under the Agreement on Transport Services in the Public Interest of rail transport, and conditions of their

Fig. 1 The territory of the Slovak Republic divided into four functional regions. *Source* Strategy for PPT by 2020 (2015)

application are subject to regulation. Fare regulation is a decision of the Transport Authority on the proposal of a railway enterprise. After approval by the party ordering transport services, it is dependent on setting the maximum for basic fares and special fares. The Transport Authority determines the fare regulation with a general binding statute.

A fare which is not subject to regulation and prices of other services provided within transport services are set by a carrier in a fare tariff according to rules for price negotiation (the prices are without the intervention/treatment/guidance of the competent authority and in accordance with the valid law on prices in the Slovak Republic). The general regulation on prices, i.e. the act No 18/1996 on prices, does not apply to regulation of fares in rail transport, but it applies to regulation of the fare in urban transport (Fig. 1) (Strategy for PPT by 2020 2015).

A fare which is not subject to regulation and prices of other services provided within transport services are set by a carrier in a fare tariff according to rules for price negotiation. The general regulation on prices, i.e. the act No 18/1996 on prices, does not apply to regulation of fares in rail transport, but it applies to regulation of the fare in urban transport.

Regional transport in the Slovak Republic is responsible for passenger trains (Os) (definition in Sect. 3.3). In the broader surrounding of the most important towns (Bratislava, Košice, Žilina, Prešov) there are also express trains "Zr" (accelerated train of the lower category and trains of REX category).

3.5 Development of Public Passenger Rail Transport in the Slovak Republic

To ensure further development of the public passenger transport system, it is necessary to identify negative aspects, bottlenecks and limitations which are related to organisation, operation and infrastructure.

Organisational: The most critical weakness of public passenger transport in the Slovak Republic is the absence of functional integrated transport systems. Currently, the Ministry of Transport and Construction of the Slovak Republic is the managing body for regional rail transport. Self-governing regions manage suburban bus transport, and relevant towns manage urban mass transport, with no mutual interconnection and coordination of individual transport modes (Strategy PPT and Strategy DTI by 2020, 2015). The timetable in rail transport is fixed with priority for long distance traffic and VÚC (autonomous region or higher territorial unit) not always able to influence the regional train times. VÚC often order bus services in partial collaboration with rail.

Operational: Rail transport has sufficient capacity to satisfy larger transport capacity across the whole network of ŽSR (Strategy for PPT and Strategy for DTI by 2020 2015). At the same time, operational costs have been increased, due to a required reconstruction of the fleet. The absence of harmonisation and integration of regional transport with other kinds of public passenger transport leads to high operational costs while part of the costs are spent on transport that is competitive for passenger rail transport. For coordination of rail and bus transport, the following is missing: change points between specific transport kinds, modern communication devices among vehicles and dispatching, and elements of transport systems integration (Strategy for PPT and Strategy for DTI by 2020 2015).

Infrastructural: Some examples of the identified infrastructural problems of regional rail transport about rail passenger transport are as follows. Insufficiently maintained and neglected character of railway stations and stops, less than optimal distribution of tariff points, infrastructure not adapted to population development and free transport in recent time, low speeds on perspective tracks which are suitable for taking line operation of their territory over (limited track speed, its transition and point restrictions) all contribute to a poor image and passenger experience (Strategy for PPT and Strategy for DTI by 2020 2015).

Identified infrastructure problems of regional rail transport for passengers include unattractive transport offerings, inconvenient travel times and long waiting times. These situations lead to passengers preferring other means of transport.

Measurements in Rail Transport for improving services

Measurements are activities by means of which strategic and specific goals are accomplished. They are fundamental for project definitions to achieve set goals. One measurement can be executed via more projects and vice versa, and one project can fulfil more measurements. The purpose of measurements are to determine the steps that must be taken at appropriate levels of the state, regional and local administration. Each measurement can also be perceived as a programme involving mutually related projects. Selected measurements focused on the support for public passenger transport are as follows (Strategy for PPT and Strategy for DTI by 2020 2015):

- To establish a department at the state level which will deal with public passenger transport in a complex manner at the conceptual and legislative level.

Currently, competencies are relatively strictly divided into road and railway transport. The potential to solve single public passenger transport modes are not utilised as organisational and basic operational parameters are not well established.

- To create transport authority (authorities) in order to cooperatively organise and order services in public interest in rail transport, suburban bus transport and urban mass transport.

The process of managing performance in regional passenger transport is only partially coordinated. In places, rail transport enters into competition with suburban bus transport (concurrency of connections) and public passenger transport is managed by too many players (the Ministry of Transport and Construction of the Slovak Republic, self-governing regions, towns and municipalities). Individual managing parties do not have the expertise for qualified transport scheduling and evaluation. Therefore their decisions go directly into hands of carriers who determine a specific form of performance order in public interest and submit it for approval to the order party.

- To establish integrated transport systems, to integrate public passenger transport in the functional region Bratislava and south-western Slovakia.

Individual subjects managing public passenger transport communicate with each other at insufficient levels which results in frequent non-coordination of rail, suburban bus and urban mass transport. Currently, there exists only one full-featured integrated transport system in the territory of the Slovak Republic (integrated transport system of Bratislava), so the advantages of a common tariff and transport operation are utilised only to a minimum extent.

- To optimise the operation of a railway network and performances in it.

Based on the analysis of transport flows, it is necessary to operate train transport on those tracks where the railway has potential and can operate the territory more efficiently than bus transport.

- To ensure the reliability of rail transport operation.

There exists a potential to tighten criteria for measuring accuracy and reliability and for creating efficient sanction mechanisms in case of non-observing defined criteria.

- To improve the maintenance of vehicles in urban and regional railway public passenger transport.

There exists a potential to reduce the break-down rate of vehicles, to increase the reliability of transport and to improve the perception of public passenger transport by passengers.

• To implement a systematic and stable schedule in regional rail transport on the railways in cooperation with the Ministry of Transport and Construction of the Slovak Republic and self-governing regions.

So far there have been frequent and vast changes in the train diagram which has resulted in reluctance and impossibility to adapt connections within suburban bus transport to trains. From the passengers' point of view, the stability of the transport system is required.

• To improve the foreknowledge of passengers and to improve the information and notification system including elements needed for passengers with impaired hearing or for those visually impaired.

No carrier in the Slovak Republic offers a special acoustic notification for blind and visually handicapped people or even orientation audio beacons at entrances into vehicles or customer services. Some carriers continue buying new vehicles without an adequate information system. (Passengers miss information about the next stop, acoustic annunciators, etc.)

• To modernise a tariff, information and communication system in railway stations, stops and trains.

The existing information systems do not present information about all available transport types of public passenger transport; the volume of provided information may be increased. Some information systems do not provide sufficient volume of online data, or they do not provide them in a sufficient time span, or with adequate accuracy and reliability.

• To procure modern train units with barrier-free access from the platform level, to procure modern locomotives with a low energy intensity, to renew the fleet for regional rail transport.

Thanks to the unification of the fleet (mainly the concentration of modern vehicles) on specific, sufficiently perspective wagon tracks, it will be possible to make use of intended infrastructural measurements in full extent.

3.6 Development of Passenger Transport in the Slovak Republic

Population mobility, expressed by some journeys or transport performance per an inhabitant/year, had a slightly decreasing tendency from 1995 to 2015 (Harmanová

Table 1 Number of passengers carried in rail transport in Slovakia for the period 1995–2017

	1995	2000	2005	2010	2015	2016	2017
Total transport of passengers (thous. person)	89,471	66,806	50,458	46,583	60,566	69,525	75,370

of which transport of passengers by operators of transport with the number of employees 20 and more

	1995	2000	2005	2010	2015	2016	2017
Total transport of passengers	89,471	66,806	50,415	46,583	60,566	69,525	75,327
International transport	3001	2474	2547	2858	3575	3718	3853
National transport	86,470	64,332	47,868	43,725	56,991	65,807	71,474

Source Statistical Office of the Slovak Republic 2018

and Štefancová 2017). Significant changes occurred in the transport division between public and non-public transport. The share of public transport on transport performance decreased from almost 50% in 1995 to 26.5% in 2015 in favour of non-public transport. The share of rail passenger transport stopped decreasing in 2013 (Harmanová and Štefancová 2017).

In the last three years (2016–2018), it increased from 1.8 to 2.3% in the case of all transport modes. The share of public rail passenger transport manifested positive development and increased from 6.2 to 8.7%. This interannual increase has been influenced by three factors, as follows (Daniš et al. 2016):

- the entry of competitors into the market,
- the introduction of free fare for selected categories of people on non-commercial trains,
- the extension of a train offer based on a higher order of train transport performances in public interest by the state.

The number of passengers carried by rail transport in the Slovak Republic in the period from 1995–2016 is shown in Table 1. A significant increase in the number of passengers carried was recorded in 2015 as a result of introducing a zero fare in rail transport (Annual Report ZSSK 2014–2017, Statistical Office of the Slovak Republic 2018).

The number of passengers carried by individual branches of public transport in the Slovak Republic is shown in Fig. 2. The introduction of free transport had a negative effect on transport performances in road transport (Annual Report ZSSK, 2014–2017, Statistical Office of the Slovak Republic 2018) (Table 2).

3.7 Public Passenger Transport Services by Rail Transport

Performance relating to the public interest include services which are the right of each citizen regardless of their financial situation and physical abilities wherever they

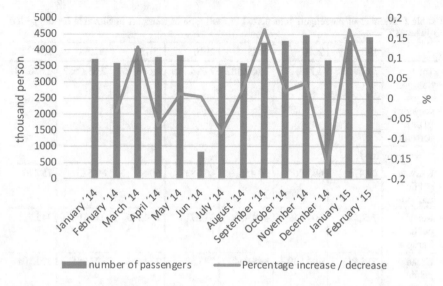

Fig. 2 Comparison of the Number of Carried Passengers before and after the Introduction of Free Transport, the Period of January 2014–February 2015. *Source* ZSSK, a.s

live. The public interest is vicariously realised through compensating the loss from the operation of non-profitable connections with the obligation to carry determined population groups in the context of relevant price legislation for free or for a set fare (i.e. social discounts) (Regulation No. 1370/2007, 2007).

The European Parliament and the Council adopted Regulation (EC) No 1370/2007 on public passenger transport services by rail and by road of 23 October 2007, repealing Council Regulations (EEC) No 1191/69 and No 1107/70 ("Regulation"). The regulation was published in the Official Journal of the European Union, L 315/1, 03 December 2007, p. 1, and it came into effect on 03 December 2009. This regulation was exerted with the Act No 488/2009 Coll, amending the Act No 168/1996 Coll on road transport subsequently amended and on amendments of some acts. This act came into effect on 03 December 2009. Currently, the issue of entering agreements on public services is treated in paragraph 21, art. 6 of the Act No 56/2012 Coll on road transport subsequently amended.

In the area of rail transport the application of the regulation has been reflected in the following acts:

- Act No 513/2009 Coll on railways and on amendments of some acts ("Act No 513/2009 Coll"),
- Act No 514/2009 Coll on railways ("Act No 514/2009 Coll").

Table 2 Number of passengers transported in each type of transport in Slovakia for the period 1995–2017

	1995	2000	2005	2010	2015	2016	2017
Total transport of passengers (thous. person)	2,669,335	2,745,442	2,669,382	2,606,149	2,608,160	2,632,582	2,630,731
of which by mode							
Railway public transport	89,471	66,806	50 458	46 583	60 566	69 525	75 370
Road public transport	722,510	604,249	449,456	312,717	252,175	259,194	245,731
Urban public transport	515,593	404,539	395,064	385,594	379,468	377,345	373,101
Inland waterway transport	138	80	134	120	132	136	12
Air transport	118	159	1716	554	583	350	411
Individual road transport	1,333,334	1,664,342	1,769,147	1,859,479	1,913,518	1,924,100	1,934,072
Public passenger transport	1,327,830	1,075,833	896,828	745,568	692,924	706,550	694,734
Non-public passenger transport	1,341,505	1,669,609	1,772,554	1,860,581	1,915,237	1,926,032	1,935,997

Source Statistical Office of the Slovak Republic 2018

3.7.1 Transport Services in the Public Interest in the Slovak Republic

Public Passenger Transport Services by Rail Transport (distant and regional) are currently ordered by the Ministry of Transport and Construction of the Slovak Republic ("Ministry") on the basis of the Agreement on Transport Services in Public Interest ("Agreement") for the period of validity 2011–2020. This Agreement is annually updated with an amendment containing the transport performance ordered for a given year where the required volume of transport performance is specified for distant and regional rail passenger transport. The ministry assigned the currently valid Agreement, which covers performances in distant rail passenger transport, directly to the selected carrier Železničná spoločnosť Slovensko, a. s. Specifically, the determination of long-distance links in the network of Železnice Slovenskej Republiky is stated in a strategic document Development Strategy of Public Passenger and Non-Motor Transport in the Slovak Republic by 2020 (Majerčák and Černá 2015).

The ministry also entered into the Agreement in case of operating passenger transport on a regional track Bratislava–Dunajská Streda–Komárno with the carrier RegioJet, a. s. The purpose of the Agreement is to preserve transport operation by public passenger rail transport by a carrier on tracks No 132 Bratislava, main station—Bratislava, Nové Mesto, and No 131 Bratislava Nové Mesto–Dunajská Streda–Komárno.

Until 2012, the market of domestic passenger transport was almost closed, and all agreements on public transport services were concluded by the ministry with the only (state-owned) carrier ZSSK. All performance of ZSSK were realised as performance in the public interest, and thus the state reimbursed a provable loss to the transport company. The situation changed only in 2012 after the new carrier RJ entered the market of domestic passenger transport. The company started to operate regional passenger transport on the line Bratislava–Komárno by a direct order and entered into an Agreement to provide services.

3.7.2 Free Transport

Pursuant to the resolution of the Government of the Slovak Republic No 530/2014, in rail passenger transport, legislative changes were introduced within measurements of a financial, economic and social package of the Government of the Slovak Republic in 2014. As an example, the introduction of free transport for selected groups of passengers on trains operated under the Agreement (Majerčák and Černá 2015).

Following the decision of the Government of the Slovak Republic, the company ZSSK, a. s., which started free transport for pupils, students and pensioners. On 22 December, the company ZSSK, a. s., and the Ministry signed an amendment to a 10-year valid agreement on operating the transport in public interest for the period 2011–2020. The scope of ordered performance in 2015 was raised to 31.304 million train kilometres. The Ministry also concluded an amendment to a 9-year valid agreement with RegioJet (Public service transport contracts in the operation of passenger transport on the track, 2018). For 2015, it ordered 1.197 million train kilometres on Bratislava–Komárno track.

Another amendment modified the formula for a new calculation of reimbursement for one train kilometre by an updated analysis of revenues and costs which would be valid from September 2015 until the end of the contractual period in December 2020.

During the first day of free transport, introduced by the company ZSSK, a. s., 104 thousand tickets were sold for that day, with 48 thousand tickets available for free transport. A significant number of paying passengers, however, bought their tickets in advance and these passengers were not included in that number. In total, the number of tickets sold was 30% higher than the number of tickets sold on Sunday 17 November 2013, a year before free transport introduction. The development of the number of passengers carried increased in this period. A more detailed overview of the number of passengers carried as well as a percentage increase/decrease within the observed months is shown in Fig. 2 (Annual Report ZSSK, 2014–2017, Statistical Office of SR 2018).

On the ground of those statistical and accounting outcomes, it is possible to state that the introduction of free transport on trains operated within the Agreement (domestic transport) led to the following development of selected indicators (Majerčák and Černá 2015):

- 18% increase in passengers,
- 11 thousand passengers more in trains under the Agreement per day,
- 42.65% non-paying passengers in domestic trains under the Agreement,
- drop in incomings in trains under the Agreement,
- an increase of 4.3 km in average transport distance on trains under the Agreement,
- an increase in seat reservation ticket sale.

The introduction of free transport led to unavoidable steps in maintaining quality standards of travelling for paying passengers, at the same time fulfilling resolutions of the Government mentioned above. Therefore, in the first phase there were set one- and more segmental limits in SC, EC, Ex and R trains. The application of limits, however, has made the travel documents issuing longer.

In the case of trains of suburban transport, there were no limits set but the capacity was increased, and transport performances in traffic peak were strengthened. As a subsequent result of the above mentioned, in the period of regular changes of the train diagram, it came to an optimisation of single trains line regarding their actual utilisation—depending on the passengers' demand for transport. At the same time, the company ZSSK operatively streamlined the capacity strength of individual trains of suburban transport.

Besides a social aspect the Government's decision to make mass transport more ecologically sound, at the same time it should evoke negotiation on rail and bus transport harmonisation. The need to harmonise schedules have been discussed for more than ten years, and there has been done almost nothing done about it. Bus carriers still act as if they are their competitors with railways and they are not willing to agree on a reasonable schedule creation. Half-empty trains and half-empty buses on the same routes next to each other are still the norm.

4   Case Study—Regional Passenger Transport in North-Eastern Slovakia

For this case study, we chose passenger transport on those regional tracks where there was low intensity of passenger transport, i.e. Humenné–Medzilaborce and

Table 3 Rail passenger transport in the direction from Humenné to Medzilaborce

No	Category of train	Number of train	HE	ML	ML–M	DMU
1	Os	8953	4:35	5:52	5:57	861
2	Os	8955	6:37	7:47	7:50	861
3	Os	8957	8:37	9:47	9:50	861
4	Os	8959	10:37	11:47	11:50	861
5	Os	8963	12:37	13:47	13:50	861
6	Os	8965	14:37	15:47	15:50	861
7	Os	8967	15:29	16:23		861
8	Interfering	8969	16:37	17:47	17:50	861
9	Os	8971	18:37	19:47	19:50	861
10	Os	8973	20:37	21:47		861
11	Os	8975	22:44	23:41		861
12	Os	8977	13:29	14:23		861
13	Os	8981	4:03	4:05		861

Legend DMU—Diesel multiple unit
Source By Schedule Booklet for the Track No 103

Humenné–Stakčín tracks. These tracks can be found in the north-eastern part of Slovakia. They serve as a good representation for studying the efficiency of rail passenger transport services in regions where there is a lack of job opportunities, educational institutions, etc. (Jaros 2018). We were looking at options to rationalise trains deployment on the basis of an analysis of their occupancy providing that transport service administration of a given region is maintained, i.e. the schedule remains unchanged.

4.1 Analysis of the Current State

To implement rationalisation measurements in the area of trains utilisation, it is necessary to start with a deep analysis of the schedule including the deployment of locomotives and their use by passengers. Tables 3 and 4 show the schedules on the track No 103 Humenné (HE)–Medzilaborce (ML) according to the current train timetable (Schedule Booklet for the Track No 103).

During a 24 h period, 12 trains travelled in the direction Humenné–Medzilaborce and ten trains in the opposite direction. The train No 8972 is an interfering train, and it overrules the ride of the train No 8962 on Sundays. All trains in this track section are trains of Os (passenger train) category; in all cases, diesel units of 861 series are planned for their transportation. The trains are organised so the connection to/from REX trains to/from Košice station and to Os trains from/to Prešov station is assured.

Table 4 Rail passenger transport in the direction from Medzilaborce to Humenné

No	Category of train	Number of train	ML-M	ML	HE	DMU
1	Os	8952	4:18	4:28	5:24	861
2	Os	8954		5:05	5:59	860
3	Os	8956	6:18	6:29	7:24	861
4	Os	8958	8:18	8:29	9:24	861
5	Os	8960	10:18	10:29	11:24	861
6	Os	8962	12:23	12:29	13:34	861
7	Os	8964	14:17	14:29	15:24	861
8	Os	8966	16:17	16:29	17:24	861
9	Os	8968	18:18	18:29	19:24	861
10	Os	8970	20:18	20:29	21:24	861
11	Loc. train	8972	12:06	12:11	13:07	861

Source By Schedule Booklet for the Track No 103

Table 5 Rail passenger transport in the direction from Humenné to Stakčín

No	Category of train	Number of trains	Departure	Arrive	DMU/Locomotion
1	Os	9401	4:37	5:24	861
2	Os	9403	5:31	6:18	757
3	Os	9405	6:37	7:24	861
4	Os	9407	8:37	9:24	861
5	Os	9409	10:37	11:24	812
6	Os	9411	12:37	13:24	861
7	Os	9413	13:37	14:24	861
8	Os	9415	14:37	15:24	757
9	Os	9417	15:37	16:24	861
10	Os	9419	16:37	17:24	861
11	Os	9421	17:37	18:24	861
12	Os	9423	18:37	19:24	861
13	Os	9425	20:37	21:24	861
14	Os	9427	22:44	23:31	861

Source By Schedule Booklet for the Track No 104

Tables 5 and 6 show the schedules on the track No 104 Humenné–Stakčín according to the current train timetable (Schedule Booklet for the Track No 104).

During a 24 h period, 14 trains travelled in the direction of Humenné–Stakčín and 15 trains in the opposite direction. The train No 9462 is an interfering train (extra train whose driving eliminates or interferes driving of regular train), and it overrules the ride of the train No 9412 on Sundays. All trains are of Os category, and they drive in an hour cycle on this track. In the case of the majority of trains, the unit of 861

Table 6 Rail passenger transport in the direction from Stakčín to Humenné

No	Category of train	Number of trains	Departure	Arrive	DMU/Locomotion
1	Os	9400	4:35	5:24	861
2	Os	9402	5:28	6:17	861
3	Os	9404	6:35	7:24	757
4	Os	9406	7:35	8:24	861
5	Os	9408	8:35	9:24	861
6	Os	9410	10:35	11:24	861
7	Os	9412	12:35	13:24	812
8	Os	9414	13:35	14:24	861
9	Os	9416	14:35	15:24	861
10	Os	9418	15:35	16:24	757
11	Os	9420	16:35	17:24	861
12	Os	9422	17:35	18:24	861
13	Os	9424	18:35	19:24	861
14	Os	9426	20:35	21:24	861
15	Os	9428	22:42	23:31	861
16	Interfering	9462	12:18	13:04	812

Source By Schedule Booklet for the Track No 104

series is used for transportation; only in the case of trains 9409 and 9412 (or in case of the interfering train 9462), a motor wagon of 812 series is used for transportation and a connection of wagons of 011 series. In the case of trains 9403, 9415, 9404 and 9418, a motor wagon of 757 series is planned to serve as a locomotive. The reason is that the principle of the running may be as follows: a train 9403 leaves Humenné for Stakčín, it returns back to Humenné as a train 9404 and continues in its drive as a train REX 1904 to Košice station. The other trains are connections of trains from/to Košice or Prešov station. Train changes happen in Humenné station which serves as a changing station.

There are three shift groups No 880, 881 and 885 for transport operation of Humenné–Stakčín and Humenné–Medzilaborce tracks. The shift group 880 operates Humenné–Prešov–Stakčín–Medzilaborce–Trebišov tracks. In this shift group, there are 5 units of 861 series set; the average daily run of a unit is 307 km. The shift group 881 operates tracks Humenné–Prešov–Bardejov–Stakčín–Medzilaborce. In this shift group, there are three units of 861 series set; the average daily run of a unit is 366 km. The shift group 885 operates tracks Humenné–Trebišov–Michaľany–Stakčín. In this shift group there are 2 motor wagons of 812 series set; the average daily run of a motor wagon is 261 km.

The analysis of a passenger frequency was conducted using an internal database of the company Železničná spoločnosť Slovensko, a. s., which regularly counts passengers in trains. Tables 7 and 8 record the occupancy of selected trains.

Table 7 Passenger frequencies in trains no. 8952 and 8953 on Humenné–Medzilaborce track

Train	Day	$\sum$ P. in a train	Average number of P. in a train	Max. number of P. in a train	Train	Day	$\sum$ P. in a train	Average number of P. in a train	Max. number of P. in a train
8952	Fri	816	49	99	8953	Fri	372	23	33
	Sat	371	22	45		Sat	224	14	16
	Sun	282	17	33		Sun	102	7	9
	Mon	709	42	93		Mon	428	26	40
	Tue	964	57	114		Tue	310	19	26
	Wed	553	33	69		Wed	295	18	22
	Thu	707	42	87		Thu	384	23	29
	Fri	453	27	60		Fri	333	20	30
	Sat	661	39	68		Sat	267	16	19
	Sun	303	18	37		Sun	131	9	12
Average		582	35	71	Average		285	18	24

Table 8 Passenger frequencies in trains no. 9400 and 9401 on Humenné–Stakčín track

Train	Day	$\sum$ P. in a train	Average number of P. in a train	Max. number of P. in a train	Train	Day	$\sum$ P. in a train	Average number of P. in a train	Max. number of P. in a train
9400	Fri	552	46	80	9401	Fri	277	24	50
	Sat	426	36	53		Sat	223	19	28
	Sun	252	22	32		Sun	95	9	13
	Mon	738	62	96		Mon	309	27	49
	Tue	702	59	98		Tue	319	27	52
	Wed	868	73	112		Wed	315	27	50
	Thu	611	51	83		Thu	296	25	54
	Fri	963	81	138		Fri	421	36	60
	Sat	304	26	39		Sat	250	22	33
	Sun	222	19	30		Sun	67	6	10
Average		564	48	77	Average		258	23	40

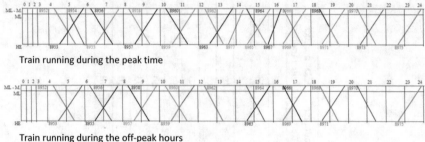

Train running during the peak time

Train running during the off-peak hours

Fig. 3 Train running on Humenné–Medzilaborce Track

Similarly, frequencies of all train connections on studied tracks were processed. Table 9 summarises the percentage occupancy of all trains on given tracks.

The analysis identified that some train connections are not well utilised in terms of capacity. During rail passenger transport operations, it is inefficient to deploy trains with motor units of a high capacity when the number of passengers is very low.

4.2 Proposal for Rationalisation Measurements

In the analysis in the previous section for the track Humenné–Medzilaborce, it was proposed to substitute a motor unit 861 (with the capacity of 179 seats) with a motor wagon 812 with a trailer 011 (with the capacity of 114 seats) at those train connections where the capacity of a DMU 861 was not utilised. A new proposal of DMU deployment is shown in Table 10.

Figure 3 shows a train running during the peak time and off-peak hours.

The total shift need for the operation of this track section during the peak time and the off-peak hours are 2 motor wagons 812 + 2 connection wagons 011 and 3 motor units of 861 series.

Table 11 shows a proposal to deploy a locomotive and DMU on Humenné–Stakčín track and Fig. 4 shows the running of the train during the peak time and the off-peak hours.

The shift need for the operation of this track section at the given frequency of trains is 1 locomotive of 757 series, 3 units of 861 series and 2 motor wagons + 2 connection wagons of 011 series. Shift needs on both tracks were calculated by the proposed running of a locomotive and DMU for weekdays and weekends, so the passenger transport during the peak time is secured.

Table 9 Train connections occupancy

Train	50–100%		21–49%		0–20%		DMU/L
	PT	OP	PT	OP	PT	OP	
Track Humenné–Stakčín							
9400			x			x	861
9401					x	x	861
9402			x				861
9403					x		757
9404	x			x			757
9405			x			x	861
9406			x			x	861
9407			x	x			861
9408			x				861
9409	x			x			812
9410			x	x			861
9411			x	x			861
9412			x	x			812
9413			x				861
9414			x				861
9415	x			x			757
9416			x	x			861
9417			x				861
9418					x		757
9419	x	x					861
9420		x	x				861
9421			x				861
9422					x		861
9423	x			x			861
9424			x	x			861
9425			x	x			861
9426			x	x			861
9427					x	x	861
9428					x	x	861
9462				x			812
Track Humenné–Medzilaborce							
8952			x			x	861
8953					x	x	861
8954			x				861
8955			x	x			861
8956	x					x	861

(continued)

Table 9 (continued)

Train	50–100%		21–49%		0–20%		DMU/L
	PT	OP	PT	OP	PT	OP	
8957			x	x			861
8958	x					x	861
8959	x			x			861
8960			x	x			861
8962	x					x	861
8963	x						861
8964			x	x			861
8965	x			x			861
8966			x	x			861
8967			x				861
8968			x	x			861
8969	x			x			861
8970					x	x	861
8971	x			x			861
8973					x		861
8975					x	x	861
8977	x						861

Legend PT—peak time; OP—off-peak hours

Table 10 A Proposal of DMU deployment in the track section Humenné–Medzilaborce

Train	PT	OP	Train	PT	OP
8952	812 + 011	812 + 011	8964	861 + 861	812 + 011
8953	861	812 + 011	8965	861	861
8954	812 + 011	–	8966	861	861
8955	861	861	8967	861	–
8956	861	812 + 011	8968	861 + 861	812 + 011
8957	861	861	8969	861	812 + 011
8958	861	861	8970	861	812 + 011
8959	861	812 + 011 + 812 + 011	8971	861	812 + 011
8960	861	861	8973	812 + 011	–
8962	861	812 + 011	8975	812 + 011	812 + 011
8963	861	–	8977	861	–

Table 11 Proposal of a locomotive and DMU deployment in the track section Humenné–Stakčín

Train	PT	OP	Train	PT	OP
9400	812 + 011	812 + 012	9415	757	861
9401	812 + 011	757	9416	861	861 + 861
9402	812 + 011	–	9417	861	–
9403	757	–	9418	757	–
9404	757	757	9419	861	861
9405	2 × 812 + 011	812 + 011	9420	861	861
9406	812 + 011	812 + 011	9421	861	–
9407	861	861	9422	861	–
9408	812 + 011	–	9423	861	861
9409	861	861	9424	861	861
9410	861	861	9425	812 + 011	812 + 011
9411	861	861	9426	861	861
9412	861	–	9427	812 + 011	812 + 012
9413	861	–	9428	812 + 011	812 + 013
9414	861	–			

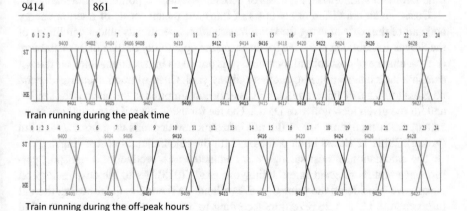

Train running during the peak time

Train running during the off-peak hours

Fig. 4 Train Running on Humenné–Statkčín Track

4.3 Economic Efficiency and Proposed Measurements

When calculating the economic efficiency, we considered the following: the number of passengers will not change, and thus the efficiency may be increased thanks to saved costs. In the case of the proposal of changing the train running there does not occur a change of all costs. Since this is a rationalisation measurement where the train arrangement is changed the indirect costs, i.e. costs of operation and administration expenses, do not change. With this proposal there is no change of costs of locomotive and train crews because it does not matter if there is a set with DMU 861 or a set

Table 12 Input parameters for costs calculation

Type of train	Gross weight in tonnes	Capacity in person	Specific energy consumption in l of PHM per thousand grtkm
861	134.16	179	13.45
861 + tow	255.16	179	13.45
812 + 011 + 812	75.00	175	11.50
812 + 011	48.00	114	11.50
812 + 011 + tow	83.00	114	11.50
757 + wagons	326.40	162	12.00

with a motor wagon 812 and trailer 011 arranged; in either case, it is necessary to have a train driver and a guard on the train. Moreover, there occurs no change of costs of using the rail infrastructure in the area of costs of ordering and assigning the capacity and the fee for organising and arranging the traffic which is dependent on train kilometres. Likewise, the costs of accessing transport points are not changed.

Costs which are different are costs of transport means, costs of traction energy and part of the costs of using the rail infrastructure. Specifically, under conditions of the Slovak Republic, it is the fee for ensuring the operability of the rail infrastructure (U_3). A change of costs of rail locomotives is not taken into account since we did not examine their dislocation into another railway yard. Costs of traction energy depend on gross tonnage, distance and specific energy consumption in the given track section and for the given locomotive or DMU. The fee for using the rail infrastructure in the area of ensuring operability of the rail infrastructure depends on gross tonnage km (grtkm). Table 12 shows input parameters for the calculation of cost savings.

The calculation of cost savings was conducted as a product of a transport performance and its relevant rate (Dolinayova et al. 2015). As to the cost savings on energy, we considered the price 0.96€/l without VAT, and the fee for using the rail infrastructure U_3 is 1.261€/grtkm according to a valid revenue (a fee for the 2nd category of a track).

In the track section Humenné–Medzilaborce costs for all train connections are the same because the train formation does not change, i.e. costs of energy are 76.2293€ and the fee U_3 is 7.1054€. In the calculation of costs per year, we separately calculated costs during a work week and during the weekend because some train connections are not established at the weekends. Cost items mentioned above were similarly calculated for Humenné–Stakčín track. Costs of a single train are as follows:

- train formation—DMU 861

  - energy—50.2421 €
  - U_3—4.5677 €

Table 13 Costs of energies and U_3 per year

	HE-ML Current State		HE-ST Current State	
	Weekdays	Weekends	Weekdays	Weekends
U3/day	156.3184022	120.7914926	154.6169602	97.86333492
Energy/day	1677.045008	1295.898415	1669.847041	1061.01275
$\sum$/day	1833.36341	1416.689908	1824.464002	1158.876084
$\sum$/year	621,260.1918		589,386.7501	
	HE-ML Proposal State		HE- ST Proposal State	
	Weekdays	Weekends	Weekdays	Weekends
U3/day	150.7764586	72.44989752	133.4356405	86.36225832
Energy/day	1606.00031	743.2898312	1429.954766	929.2948379
$\sum$/day	1756.776768	815.7397287	1563.390406	1015.657096
$\sum$/year	533,004.2608		507,648.1676	

- train formation Locomotion 757 + wagons

  – energy—116.564 €
  – U_3—11.1129 €

- train formation—DMU 812

  – energy—24.012 €
  – U_3—2.5535 €

Cost savings quantification comes out of the current and proposed deployment of a locomotive and DMU in Tables 10 and 11.

Costs at the current state and in our proposal can be found in Table 13—they are listed separately for weekdays and weekends about a different number of connections. The calculation at the proposed state considers an optimal running of a locomotive and DMU shown in Figs. 3 and 4.

Total costs of the fee U_3 and energies at the current state on both tracks represent the sum 1,210,646.942 €. In our proposal, the amount of these costs is 1,040,652.428 € which represents the total difference of 169,994.514 €/year.

5 Conclusion

The future growth of the rail transport share on the passenger transport service market is based on a combination of: competition on a commercial principle, orders for public services, and regulatory measurements.

The economic efficiency of short-distance rail passenger transport is impacted by many factors. Some of them are of a social or economic character (such as the obligation to ensure transport service administration of a region, economic level of a region, etc.) which a carrier cannot influence. This study shows that an optimal setting of processes may bring cost savings to rail transport operators which may ultimately increase the economic efficiency while preserving quality standards of public passenger rail transport. Though, the Government's interventions on prices may eliminate the benefits of new competitors' entry as well as realised investments into the infrastructure development and purchase of mobile resources for passenger transport (Regulation No. 1370/2007, 2007).

Acknowledgements The paper was supported by the VEGA Agency, Grant No. 1/0019/17 "Evaluation of regional rail transport in the context of regional economic potential with a view to effective use of public resources and social costs of transport", at Faculty of Operations and Economics of Transport and Communication, University of Žilina, Slovakia.

References

Annual Report ZSSK, a.s. 2014–2017, 10/2018. [online] Available at: http://www.slovakrail.sk/sk/o-spolocnosti/vyrocne-spravy0.html

Avenali A, Boitani A, Catalano G, D'Alfonso T, Matteucci G (2018) Assessing standard costs in local public bus transport: a hybrid cost model. Transport Policy (62). Special Issue:48–57

Bogart D, Chaudhary L (2015) Off the rails: Is state ownership bad for productivity? J Comp Econ 43(4):997–1013

Cantos P, Pastor JM, Serrano L (2012) Evaluating European railway deregulation using different approaches. Transport Policy 24:67–72

Catalano G, Daraio C, Diana M, Gregori M, Matteucci G (2019) Efficiency, effectiveness, and impacts assessment in the rail transport sector: a state-of-the-art critical analysis of current research. Int Trans Oper Res 26(1):5–40

Daniš J, Dolinayová A, Černá L (2016) Direction of Slovak transport strategy. Logistics–Economics–Practice. Available at: http://www.logistickymonitor.sk/

Dolinayova A, Danis J, Cerna L (2017) Regional railways transport-effectiveness of the regional railway line. Sustainable rail tra nsport, vol 2. In: Conference on sustainable rail transport (Rail-Newcastle) 2017, pp 181–200. Springer

Dolinayova A, Loch M, Kanis J (2015) Modelling the influence of wagon technical parameters on variable costs in rail freight transport. *Research in Transportation Economics* Vol. 54 (Dec. 2015) p. 33–40

Fitzová H, Matulová M, Tomeš Z (2018) Determinants of urban public transport efficiency: case study of the Czech Republic. Eur Transp Res Rev 10(2). Article number 42

Fourth railway package (2016) [online] Available at: https://www.consilium.europa.eu/sk/policies/4th-railway-package/

Fraszczyk A, Lamb T, Marinov M (2016) Are railways really that bad? An evaluation of rail systems performance in Europe with a focus on passenger rail. Transp Res. Part A-Policy Pract 94:573–591

Gasparik J, Abramovic B, Zitricky V (2018) Research on dependences of railway infrastructure capacity. Tehnicki Vjesnik-Technical Gazette. 25(4):1190–1195

Growitsch C, Wetzel H (2009) Testing for economies of scope in European Railways. An efficiency analysis. J Transport Econ Policy 43:1–24

Harmanová D, Štefancová V (2017) European transport policies for the development of railway transport. Railway transport and logistics. J Rail Transport, Logistics Manage 13(2) [online] Available at: https://zdal.uniza.sk/images/zdal/archiv/zdal_2017_02.pdf

Jaros M (2018) Support of the railway passenger transport rationalization in the Humenné–Stakčín and Humenné–Medzilaborce tracks. Diploma thesis, University of Žilina

Majerčák J, Černá L (2015) Registration accounts for free shipping. Logistics–Economics–Practice

Marinov M, Sahin I, Ricci S, Vasic-Franklin G (2013) Railway operations, time-tabling and control. Res Transp Econ 41(1):59–75

Masek J, Kendra M, Milinkovic S, Veskovic S, Barta D (2015) Proposal and application of methodology of revitalisation of regional railway track in Slovakia and Serbia. Part 1: theoretical approach and proposal of methodology for revitalisation of regional railways. Transport Prob 10:85–95

Panak M, Nedeliakova E, Abramovic B, Sipus D (2017) Synergies of the liberalization of the railway transport market. In. MATEC web of conferences, vol 134, Article Number 00045

Public service transport contracts in the operation of passenger transport on the track, 10/2018. [online] Available at: https://www.mindop.sk/ministerstvo-1/doprava-3/zeleznicna-doprava/zmluvy-o-dopravnych-sluzbach-vo-verejnom-zaujme-pri-prevadzkovani-osobnej-dopravy-na-drahe

Regulation (EC) No. No 1370/2007 of the European Parliament and of the Council of 23 October 2007 on public passenger transport services by rail and by road and repealing Council Regulations (EEC) (EEC) No 1191/69 and (EEC) 1107/70, 10/2018 [online] Available at: https://publications.europa.eu/sk/publication-detail/-/publication/b363bd7c-700b-4360-a9af-82156c6be71a

Rüger B (2017) Influence of passenger behaviour on railway-station infrastructure, sustainable rail transport, vol 2. In: Proceedings of RailNewcastle, pp 127–160. Springer

Schedule Booklet for the Track No 103 Łupków PL–Medzilaborce–Michaľany, No. 1297/2016–O 410. 2016. Bratislava. Železnice Slovenskej republiky

Schedule Booklet for the Track No 104 Stakčín–Humenné, No. 1297/2005–O 410. 2014. Bratislava. Železnice Slovenskej republiky

Singhania V, Marinov M (2016) An event-based simulation model for analysing the utilization levels of a railway line in urban area. PROMET, 29(5):521–528

Smith ASJ, Benedetto V, Nash C (2018) The impact of economic regulation on the efficiency of European railway systems. J Transport Econ Policy 52, Part 2:113–136

Statistical Office of the Slovak Republic (2018) 10/2018. [online] Available at: www.slovak.statistics.sk

Stoilova S (2018) Study of railway passenger transport in the European Union. Tehnicki Vjesnik-Technical Gazette 25(2):587–595

Strategy for DTI by 2020. Strategic plan for development of transport infrastructure of the Slovak republic by 2020, 10/2018 [online] Available at: www.mindop.sk/ministerstvo-1/doprava-3/strategia/strategicky-plan-rozvoja-dopravnej-infrastruktury-sr-do-roku-2020

Strategy for PPT by 2020 (2015) Strategy for the development of public passenger and non-motor transport by 2020, strategic document of the Slovak Republic, 11/2018. [online] Available at: https://www.mindop.sk/ministerstvo-1/doprava-3/verejna-osobna-doprava/strategicke-dokumenty/strategia-rozvoja-verejnej-osobnej-dopravy-sr-do-roku-2020

Weerawat W, ThongboonpianT, Fraszczyk A (2017) Impact of a "Missing Link" on passenger's travel time: a case study of Tao Poon station. Sustainable rail transport, vol 2. In: Conference on sustainable rail transport (RailNewcastle), pp 81–92. Springer

White Paper (2011) Roadmap to a single European transport area, 10/2018 [online] Available at: https://eur-lex.europa.eu/legal-content/SK/TXT/?uri=CELEX%3A52011AR0101

Sustainable Railway Solutions Using Goal Programming

Pedro Henrique Del Caro Daher, Diogo Furtado de Moura,
Gregório Coelho de Morais Neto, Marta Monteiro da Costa Cruz
and Patrícia Alcântara Cardoso

Abstract Project selection and the prioritization of activities are configured as classical optimization problems, and one of the most commonly used techniques to solve this kind of problem is Goal Programming (GP), a multi-criteria analysis technique. Problems related to the prioritization of railway investment selection or maintenance processes involve goals and constraints such as budget constraints, the availability of labour and resources, and the degradation of permanent track materials. This chapter presents two applications of Goal Programming in railways. The first model selects projects for a railway-sustaining investment portfolio. The second model allows for the prioritization of railway superstructure maintenance based on the maintenance demand for components and the geometric, environmental and demographic characteristics of a railway. Defining the best investment portfolio or a proper maintenance strategy are essential tasks for railway sustainability to achieve long-term goals involving multiple, often immeasurable and conflicting, objectives. The results show that these two proposed models allow for the prioritization of goals defined as the most important and proved useful in the presentation of scenarios that facilitate the choice of investment portfolio or superstructure maintenance strategy.

Keywords Project investments · Railway maintenance · Goal programming

1 Introduction

According to (Vargas 2010), one of the main challenges of organizations lies in their ability to make certain and consistent choices in line with their strategic direction. Thus, the contribution of investment funds to projects that bring positive returns becomes fundamental for the company that wants to succeed in its business and achieve its long-term goals. However, it is essential to use strategic planning to

P. H. Del Caro Daher · D. F. de Moura · G. C. de Morais Neto · M. M. da Costa Cruz (✉) ·
P. A. Cardoso
Civil Engineering Graduate Course, Federal University of Espirito Santo,
Vitória, Brazil
e-mail: marta.cruz@ufes.br

© Springer Nature Switzerland AG 2020   129
M. Marinov and J. Piip (eds.), *Sustainable Rail Transport*, Lecture Notes in Mobility,
https://doi.org/10.1007/978-3-030-19519-9_4

effectively allocate their resources. Strategic planning provides an environmental analysis of a company, identifying its opportunities, threats, strengths and weaknesses to get out of their current state (mission) and reach their expected state (vision) (Valle et al. 2007). Additionally, railway maintenance can be considered a strategic activity since the results are directly related to performance, reliability, transport safety and cost reduction. The maintenance activity is an important part of total railway costs.

The operation of a railway requires financial resources that can be divided into two categories: CAPEX (*Capital Expenditure*) and OPEX (*Operational Expenditure*). The CAPEX means that resources are used for investments that will bring future benefits. Within this last category, there are current investments that aim to maintain or increase the productivity of assets, improve the quality of products/services provided, preserve the environment and/or work conditions or even meet requirements imposed on the railway by external and internal authorities. The OPEX means that resources are used for the maintenance of assets and for the payment of staff.

Once the criteria for financial allocation resources have been defined, optimization models are excellent tools to help in making decisions. According to (Ahern and Anandarajah 2007), these models are developed to help decision making and select projects identified as possible investments. The selection of projects of railway investments presents itself as a problem of multiple objectives, and the development of a model for this purpose will use Goal Programming (GP) as a methodology. GP is a technique in which one or more goals are formulated as constraints, having an objective function that seeks to minimize the sum of the absolute deviations of these goals (Ahern and Anandarajah 2007). The same condition is found in railway maintenance. According to (Ferreira 2010), with robustness and high investment involved in the maintenance process, railways start to adopt maintenance investment strategies increasingly directed to critical problems that have been identified. However, a misconception in the railway maintenance process can result in defects concentrated in weak points that, if not properly identified and repaired, can create permanent and irreversible deformations in the railway, with the solution requiring complete replacement of components of the whole stretch with possible unwanted interruptions in train movements.

One of the ways to simplify decision making is through some method of optimization. The technique proposed by Charnes and Cooper (1961) to facilitate this process is Goal Programming (GP), an area of multi-objective programming that treats possible restrictions related to a problem that can be considered flexible to allow values close to those previously established as goals. This chapter will present two applications of the GP technique in railways. First, this chapter will present basic concepts of GP used to develop these two models, and in sequence, the applications will be described. The models were developed using the Solver Supplement of Microsoft Excel.

2 The Goal Programming

With the complexity existing in today's organizations, decision makers try to max-imize generally not well-defined execution functions. The conflicts of interest and the lack of complete information make it virtually impossible to construct a reliable mathematical function that better represents decision-maker preferences. Thus, with the absence of an ideal decision-making environment, the decision maker tries and achieves a series of goals (or targets) as close as possible (Tamiz and Jones 1998).

Decision-making models with multiple objectives and goal programming (GP) are important tools for the fields of operational research and other management sciences, with extensive application in engineering and science investigations. The complexity that exists in most of the real problems is due to difficulties in modelling and solving problems with a single objective. GP is a method that optimizes multiple goals, minimizing the deviations of the objectives of the aspiration levels or goals sets by the decision maker. Deviations near zero show that the targets have been achieved. These deviations can also be positive or negative, which means that the targets were reached below or above a defined target (Colapinto et al. 2015).

The first application of GP was made by Charnes et al. (1955) and Charnes and Cooper (1961) in the context of executive remuneration. At that time, the term GP was not used, and the model was an adaptation of linear programming (Tamiz and Jones 1998). Since then, the GP became one of the methods of optimizing multiple goals that was more used considering the evolution of the number of articles pub-lished (Colapinto et al. 2015). The main important methods of GP are weighted GP, which allows for a trade-off analysis between the unwanted deviation variables, and lexicographic GP, which has different levels of priorities, each one containing several unwanted deviation variables to be minimized. In the sequence, the terminology and structure of weighted and lexicographic GP will be presented.

2.1 GP Basic Concepts and Classification

The concept of objectives and attributes is essential to any decision-making process. According to (Morais Neto 1988), an objective is essentially an expression that reflects the will of the decision-maker about a certain state of the system in question. An objective is an expression of decision-maker desire, and thus, it can be fully achieved or not. The author cites a decision problem with multiple objectives that is characterized by several goals, some well-defined, others poorly defined. A set of well-defined objectives often presents a hierarchical structure similar to the structure shown in Fig. 1.

The terminology used in the GP varies widely in the literature. Among the various terms used in the GP issues, the basic definitions cited by (Morais Neto 1988) are as follows:

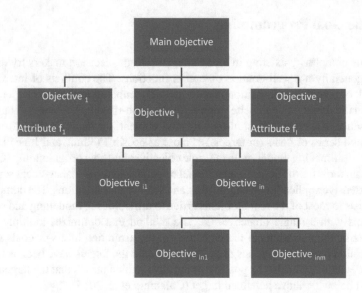

Fig. 1 Objective hierarchy. *Source* Morais Neto (1988)

- Objective: The expression (in a narrative or quantitative way) that reflects decision-maker desire, such as maximize profits or minimize costs.
- Level of achievement: A specific value that is associated with a desirable or affordable level of achievement of an objective, which is used as a measure for goal achievement;
- Goal: Every objective that has an achievement level is called a goal. Therefore, to reach a profit of at least $Y or to reduce costs, a maximum of $X are examples of goals;
- Goal deviation: The difference that may occur between the level of achievement attained for a goal and the level of achievement initially desired (aspiration level). A goal deviation can be "more" or "less" in relation to the aspiration level. In other words, you can have positive or negative deviations that occur when the goal is not reached.

According to (Tamiz and Jones 1998), GP models can be classified into two macro-groups. In the first group, the weighted GP, deviation variables receive weights according to their importance as established by the decision maker and have the following objective function:

$$\min Z = \sum_{i=1}^{I} (u_i n_i + v_i p_i) \qquad (1)$$

Subject to:

$$f_i(x) + n_i - p_i = b_i, \quad i = 1\dots I \tag{2}$$

when

$f_i(x)$	linear function of x, for objective i;
x	decision variable;
b_i	desired value to reach each goal;
n_i and p_i	negative and positive deviation variables for a target value;
u_i and v_i	weights associated for deviation variable in Z function; and
I	total number of objectives.

The second group is classified as lexicographic GP. This group is based on optimizing goals according to their relative importance to the decision maker. The most important goals will be at the highest priority level, while the less important goals will be at the lowest levels. The deviation values obtained from a high priority level will be considered restrictions to its lower priority levels. In other words, the objectives of lower priority levels will play a secondary role in the decision-making process Aouni et al. (2014). The lexicographic GP is represented algebraically by the following objective function, and it is subject to the same previous restrictions (Morais Neto 1988):

$$Lex \min \ a = [g_1(n_i, p_i), (g_2(n_i, p_i), \dots, g_L(n_i, p_i)] \tag{3}$$

subject to:

$$f_i(x) + n_i - p_i = b_i, \quad i = 1\dots I \tag{4}$$

where:

$f_i(x)$	linear function of x, for objective i;
x	decision variable;
a	ordered vector of priority levels;
b_i	desired value to reach each goal;
$g_L(ni, pi)$	linear function of deviation variables for priority level L;
L	priority levels; and
n_i and p_i	negative and positive deviation variables for a target value;
I	total number of objectives.

In summary, the main difference between weighted and lexicographic GP methods is that in the lexicographic method, the goals are optimized in sequence, according to the priority level that the decision-maker sets for each one of the goals. That is, the lexicographic GP model will only seek the optimization of the second priority goal after optimizing the first priority and so on. In weighted GP, the model is optimized, seeking the best result according to the objective function, respecting the weights assigned to the deviation variables of each objective. The decision maker may work with different trade-offs and can generate an analysis of different scenarios and choose the best to meet his/her needs.

The literature discusses the normalization techniques in GP problems several times, aiming to overcome the use of incommensurability, which occurs when goals have different units of measure. In these cases, the sum of deviation goals in an objective function could provoke incorrect results. The most popular normalization techniques are Euclidean normalization, scaling between ideal and nadir points and percentage normalization (Tamiz et al. 1995). The Percentage Normalization will be used in this chapter. This technique considers that each deviation from goal is divided by its goal value and then multiplied by 100. Then, these deviations come to represent the goal deviation percentage. The critical factor of this approach is the goal target value because the method only works well when target values are not equal to zero.

2.2 The Use of GP in Transports

According to Jones and Tamiz (2010), there is a tendency to use weighted GP rather than lexicographic GP because weighted GP provides greater flexibility by weighting constants and by the desire of decision makers to create more analysis of trade-offs and comparisons between the objectives. Niemeier et al. (1995) developed five optimization models for the selection of a set of projects with the objective of improving the performance of an entire hypothetical transport system. Among the models, one model used the weighted GP. Uliana (2010) used weighted GP associated with the utility method to solve the distribution problem of natural gas using trucks and/or pipelines. Yang et al. (2011) also used weighted GP to develop a freight optimization model of a Chinese intermodal network for the Indian Ocean, seeking to minimize transport costs, transit time and variation, ensuring a continuous flow and compatibility among railways, freeways, ships, airplanes and non-oceanic waterways.

Lexicographic GP (LGP) was used by (Morais Neto 1988) to develop a model of allocation of military cargo flows, seeking to rationalize the work of the military planner and the use of the available transport subsystems. Ramos (1995) applied LGP in deciding which mitigation alternatives or actions should be adopted in marine oil terminals to improve their operational performance. LGP was also used to develop a network design model for expanding a railway rapid transit network with a given number of new lines (López-Ramos et al. 2017).

Decisions considering transport investment projects were studied by Teng and Tzeng (1996) who developed a method of selecting independent transport investment alternatives through an efficient distance heuristic algorithm, seeking to maximize the objectives achieved, according to available resources. Additionally, Ahern and Anandarajah (2007) developed a weighted integer GP (WIGP) model for selection of new railroad investment projects in Ireland, where the core business is passenger transport. Subsequently, Ahern and Anandarajah (2008) developed a quadratic model that was applied in a similar situation, selecting new railway projects for passenger trains in Ireland, using the ideal solution concept, allowing more than one optimal solution to be identified as the first and the second and, thus, a potentially better solution, whereas the GP model presented only one optimal solution.

Some authors considered that usual GP methods for transport investment prioritization projects are not able to effectively deal with preferences and uncertainties of decision makers; therefore, they proposed the use of fuzzy set theory to deal with inaccurate information. As an example, Teng and Tzeng (1996) developed a method of selecting independent transport investment alternatives through an efficient distance heuristic algorithm, seeking to maximize objectives achieved, according to available resources. Teng and Tzeng (1998) also applied this theory and proposed a fuzzy GP 0-1 model, which was applied in a hypothetical situation for the selection of transport investment projects, considering 10 projects with resource constraints and objectives to be reached, which were qualitative and quantitative targets to be achieved. Kahraman and Büyüközkan (2008) combined the fuzzy GP with the fuzzy Analytic Hierarchy Process (AHP) to prioritize projects using the Six Sigma methodology. Chang et al. (2009) proposed an integrated model for selection of revitalization projects of the Alishan Forest Railway in Taiwan based on the fuzzy Delphi and Analytic Network Process (ANP) for qualitative evaluation of prioritization criteria of these projects. The results of this evaluation were incorporated into a GP 0-1 model to aid decision making. Wey and Wu (2007) proposed a methodology for the selection of transport infrastructure projects, combining the Delphi fuzzy method with ANP and GP 0-1. The authors applied the model to study transport infrastructure improvement in Taichung City, Taiwan.

This section presented some basic concepts and applications of GP in transport. In the next section, we present two applications of weighted GP in railways, referring to applications in forecasting new investments and in maintaining railway assets.

3 Applications

Weighted GP modelling was used for both: railway project selection and railway superstructure maintenance selection. These two case studies were validated using a heavy haul railway in Brazil.

3.1 Railway Project Selection

The general objective for this problem is to prioritize of current investment projects of a railway, according to strategic planning, financial indicators and sustainability. This objective needs to be decomposed into specific objectives to identify their attributes and measurement units, as described in Tables 1, 2 and 3, as follows.

Table 1 Specific objective details

ID	Goal	Attribute	Unit
1.1	To ensure projects that do not exceed the financial resources available for investment	Resources	R$*
1.2	To ensure financially viable projects (net present value)	VPL	R$
1.3	To ensure projects that contribute physical availability (PA) of railway increase	DF	%
1.4	To ensure projects that guarantee minimum conditions for sustainability requirements of the business	–	–

Note R$—Brazilian Real

Table 2 Decomposition of 1.4 objective

ID	Goal	Attribute	Unit
1.4.1	To ensure projects that contribute to reducing the rate of accidents (AC)	Accident rate decrease	Accident/MTkm[a]
1.4.2	To ensure projects that meet the expectations of external stakeholders	–	–

[a]Million tons per kilometres

Table 3 Attributes and Units – external stakeholders

ID	Goal	Attribute	Unit
1.4.2	To ensure projects that meet the expectations of external stakeholders		U_i
	– To meet environmental requirements	Environmental standards	U_{pj}
	– To meet ANTT requirements	ANTT Commitments	U_{pj}
	– To meet local community requirements	Commitments Communities	U_{pj}

The specific objective 1.4, listed above in Table 1, has neither an attribute association nor measures because it is described in a generic way. Therefore, this objective was decomposed to more specific objectives, as presented in Table 2.

Even after the second decomposition of objectives, objective 1.4.2 is still without an attribute and associated measure. Thus, for this objective, the following subobjectives can be identified, among others: (a) to comply with environmental requirements, (b) to comply with ANTT[1] requirements, and (c) to meet requirements of local communities. The association of attributes and measurement units in each subobjective uses an associated utility method to ensure that utilities of each attribute are mea-

---

[1]Brazilian Transport Agency (Agência Nacional de Transportes Terrestres – ANTT).

sured on a single numerical scale. In this case, quantification is performed by the association of an abstract value of usefulness for each possible situation. Therefore, an event that has no numeric or monetary correspondent can be transformed into a utility value (Margueron 2003). Table 3 shows the resulting attributes and units for subobjective 1.4.2:

Thus, the associated utility of external stakeholder objectives for each project can be calculated by the formula, and is defined as "Attendance to stakeholders":

$$U_p = \sum_{j=1}^{J} w_j u_{pj} \tag{5}$$

where

U_p total utility of p project ($p = 1,..., P$);
w_j associated weight of attribute j ($j = 1,..., J$); and
u_{pj} utility of attribute j.

The goal is achieved by the inclusion of negative and positive deviations in objective mathematical expressions and the attribution of its target or level of achievement. For the problem in question, the goals are:

$$\text{Budget} - \text{O} \quad (\sum_{p=1}^{P} X_P C_P) + n_i - p_i = M_O \tag{6}$$

$$\text{Net present value} - \text{NPL} \quad (\sum_{p=1}^{P} X_P VPL_P) + n_i - p_i = M_{VPL} \tag{7}$$

$$\text{Physical Availability} - \text{DF} \quad (\sum_{p=1}^{P} X_P DF_P) + n_i - p_i = M_{DF} \tag{8}$$

$$\text{Accident rate decrease} - \text{AC} \quad (\sum_{p=1}^{P} X_P AC_P) + n_i - p_i = M_{AC} \tag{9}$$

$$\text{Attendance to stakeholders} - \text{U} \quad (\sum_{p=1}^{P} X_P U_P) + n_i - p_i = M_U \tag{10}$$

where

x_i Binary variable for selection of projects ($0 = $ no; $1 = $ yes) for p projects ($p = 1, ...P$);
n_i and p_i Negative and positive deviation variables of goal i ($i = O, VPL, DF, AC$ or U);
C_p Investment required for project p implementation [R$];
VPL_p Net Present Value of project p [R$];
DF_p Impact on physical availability of the railway caused by project p [%];

AC_p Annual reduction rate of railway accidents caused by the project p. [A/MTkm];

U_i Total utility of project p;

M_i Target values for each goal i ($i = O, VPL, DF, AC\ or\ U$); and

P Number of projects considered.

3.1.1 Model Formulation

Once the goals are defined, an objective function is proposed that will seek to minimize the weighted sum of deviation variable percentage of goal values defined for each goal. Thus, the decision maker can prioritize the goals, normalized by percentage where deviations are lower for the most important goals. Thus, the model is composed of the execution function, and goals are defined as follows:

$$Minz = \sum_{i=1}^{I} \left(u_i \frac{n_i}{M_i} + v_i \frac{p_i}{M_i} \right) \tag{11}$$

Subject to:

$$\left(\sum_{p=1}^{P} X_P C_P \right) + n_i - p_i = M_O \tag{12}$$

$$\left(\sum_{p=1}^{P} X_P VPL_P \right) + n_i - p_i = M_{VPL} \tag{13}$$

$$\left(\sum_{p=1}^{P} X_P DF_P \right) + n_i - p_i = M_{DF} \tag{14}$$

$$\left(\sum_{p=1}^{P} X_P AC_P \right) + n_i - p_i = M_{AC} \tag{15}$$

$$\left(\sum_{p=1}^{P} X_P U_P \right) + n_i - p_i = M_U \tag{16}$$

where

$n_i, p_i \geq 0$ for all i

$M_i > 0$ for all i

x_p must be a binary value

$0 < u_i < 10$

$0 < v_i < 10$

u_i and v_i: positive and negative weighting deviations for goal i

M_i: Target value for the goal i.

3.1.2 Model Application

The model for selection of railway private company investment projects was used for a portfolio selection of 15 fictitious projects considering values equivalent to a real case, using the solver supplement of *Microsoft Excel* 2013®. The 15 fictitious projects considered for this application has the purpose to ensure the confidentiality information from real projects. The data used do not influence the application and results of the model. The total cost of all 15 projects is R$2418 $\times$ 10^6, and the available budget for project portfolio selection is equal to R$1800 $\times$ 10^6 (74.4% of project total cost). According to the decision maker strategy, the target values of the other goals will be equal to 85% of the total value of each attribute of the portfolio, as shown in Table 4. The budget was considered a goal and not a rigid constraint, so the decision maker can work with different trade-offs from a potential increase in the available budget. To analyse the different trade-offs several scenarios will be generated, changing the weight of each objective, so the decision maker is able to compare the projects selected by the model in each situation. The scenarios will be explained below. Thus, the GP model will indicate which of 15 projects will be selected, seeking to minimize the percentage deviation of target values per goal from decision variable X_i, which will be 1, if project i was selected and 0 otherwise. The values presented in Table 4 are equivalent to a real case.

According to the railway strategic planning the objectives considered in the project selection model have the following priorities: Budget (O), Net Present Value (VPL), Accident Rate Decrease (AC), Attendance to stakeholder (U) and physical availability (DF), as presented in Fig. 2.

It is noteworthy that the model will compose the portfolio of projects considering that the higher the value of the weights (v_i and u_i) of deviation variables of a goal, the higher the level of its priority because the objective function of the model will seek to minimize the weighted sum of the percentage deviation variables of the target values. With these assumptions, for the first scenario, the model selected 11 projects among the 15 proposed. Analysing Table 5, we observe that the budget objective was the only one with a non-zero v_i value. The other u_i variable objectives received positive values since a burst in the available budget would not be acceptable by decision

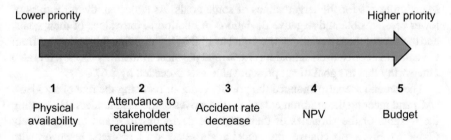

Fig. 2 Order of objective priority

Table 4 Railway projects portfolio

Project	Selector (Xi)	Budget (R$ × 10^6)	NPV (R$ × 10^6)	DF (%)	Accident rate decrease (A/MTkm)	Attendance to Stake-holder (U)
1		207	2557	0.206	0.000	3.000
2		132	0477	0.218	0.075	2.333
3		180	1942	0.436	0.009	2.000
4		199	2145	0.515	0.005	3.667
5		120	0921	0.060	0.011	3.667
6		156	1868	0.988	0.000	3.000
7		153	0612	0.735	0.001	2.667
8		189	2952	0.272	0.015	2.667
9		128	0444	0.220	0.031	1.667
10		180	0710	0.778	0.010	3.333
11		165	0037	0.007	0.000	2.000
12		150	2298	0.197	0.059	3.667
13		174	2324	0.487	0.008	2.333
14		105	1240	0.429	0.022	3.000
15		180	0533	0.057	0.004	2.333
Total		2418	21,060	5.605	0.250	41.333
Target value		1800	17,901	4.764	0.212	35.133
Target value/Total (%)		74.4	85	85	85	85

maker. For the other goals, the higher the values achieved the better, even if they eventually exceed their target values.

In this scenario, the objective function reached 35.6%, which shows that there may be an imbalance between the weights of the deviation variables or very high selection levels for the target values of some goals. As expected, the objectives of lower priority obtained negative deviations in relation to the values of their goals, and the sum of these deviations was equal to 11.2%, highlighting the deviations from the target of service to stakeholders (7.0%) and physical availability (3.7%). It is also noteworthy that the goal of net present value was exceeded by 8.6%.

The second scenario changed the priority order of reducing the rate of accidents (AC) and meeting the requirements of the objectives of the stakeholders (U), passing the weights of the variables of their negative deviations to 4 and 6, respectively (Table 6). From this change, the model again selected 11 projects; however, with the alteration of a project in relation to the initial situation, presenting an objective function equal to 42.3% and the following values and deviations was achieved.

Table 5 First scenario results

Objective	Values			Deviations		Weights	
	Target	Reached	Difference	ni (%)	pi (%)	ui	vi
O	1800.000	1792.000	−8.000	0.4	0.0	0	10
NPV	17.901	19.433	1.533	0.0	8.6	8	0
DF	4.764	4.586	−0.179	3.7	0.0	2	0
AC	0.212	0.214	0.001	0.0	0.7	6	0
U	35.133	32.667	−2.467	7.0	0.0	4	0
Total				11.2	9.2		

where O = Budget, NPV = Net Present Value, DF = Physical Availability, AC = Accident Rate, and U = Utility function for stakeholder requirements

Table 6 Second scenario results

Objective	Values			Deviations		Weights	
	Target	Reached	Difference	ni (%)	pi (%)	ui	vi
O	1800.000	1765.000	−35.000	1.9	0.0	0	10
NPV	17.901	18.104	0.203	0.0	1.1	8	0
DF	4.764	4.884	0.120	0.0	2.5	2	0
AC	0.212	0.206	−0.006	2.9	0.0	4	0
U	35.133	33.333	−1.800	5.1	0.0	6	0
Total				10.0	3.7		

where O = Budget, NPV = Net Present Value, DF = Physical Availability, AC = Accident Rate Decrease, and U = Utility function for Attendance to stakeholders

In this second scenario, the total value of the selected portfolio was R$27 million lower than in the previous one, and the sum of the negative deviations totalled 10.0%, which is 1.2% lower than the previous situation. If we compare the sums of negative deviations between the two scenarios, disregarding the budget deviations, there was a reduction of 2.7%, which explains the increase in the negative deviation of the budget target. Moreover, to compensate for this reduction, the positive deviations from the other targets were reduced from a total of 9.2 to 3.7%.

To evaluate the necessary increase in the budget so that the negative deviations of the targets were the smallest possible, one can opt for a new scenario, reducing the priority of the budget target and maintaining the priority of the first scenario, which reflects the priorities of objectives according to the strategic planning of the organization. The budget is still a priority but is less important since the weight of its deviation variable received the smallest nonzero value in relation to the other goals. Thus, for the third scenario, the model selected 12 projects, with a burst of R$93 million in the budget target, but with the negative deviation of only 0.4% in the target of meeting the requirements of the stakeholders and with objective function equal to 12.6%. The other goals had positive deviations, as shown in Table 7.

Table 7 Third scenario results

Objective	Values			Deviations		Weights	
	Target	Reached	Difference	ni (%)	pi (%)	ui	vi
O	1800.000	1893.000	93.000	0.0	5.2	0	2
NPV	17.901	18.548	0.647	0.0	3.6	10	0
DF	4.764	5.105	0.340	0.0	7.1	4	0
AC	0.212	0.237	0.025	0.0	11.7	8	0
U	35.133	35.000	−0.133	0.4	0.0	6	0
Total				0.4	27.6		

where O = Budget, NPV = Net Present Value, DF = Physical Availability, AC = Accident Rate Decrease, and U Attendance to stakeholders

Table 8 Fourth scenario results

Objective	Values			Deviations		Weights	
	Target	Reached	Difference	ni (%)	pi (%)	ui	vi
O	1800.000	2083.000	283.000	0.0	15.7	0	0
NPV	17.901	18.059	0.158	0.0	0.9	10	0
DF	4.764	5.178	0.414	0.0	8.7	4	0
AC	0.212	0.219	0.006	0.0	3.0	8	0
U	35.133	36.667	1.533	0.0	4.4	6	0
Total				0.0	32.7		

where O = Budget, NPV = Net Present Value, DF = Physical Availability, AC = Accident Rate Decrease, and U = Attendance to stakeholders

Even with a significant reduction of negative deviations in third scenario, how much more budget would be needed for all negative deviations to be equal to zero? This would mean that all target values of the objectives were achieved or were larger than that established by the decision maker. To evaluate this trade-off, a new scenario was carried out, with the weights of the deviation variables maintained in relation to the previous situation, except for the positive deviation variable of the budget, which went from 2 to zero, showing that this deviation ceased to be a priority and became indifferent to the decider. The results of this scenario will consider no budget restriction and will show how much more budget would be necessary, so the decision maker could reach at least the target level of all the other objectives. In these conditions of the fourth scenario, the model selected 13 projects, with an overflow of R$283 million in the budget target, and an objective function equal to 0.0% and no negative deviation, as shown in Table 8.

In the analysis of the results in the first scenario (Table 5), priorities were defined according to the strategic planning and available budget, and from the results achieved, there were reversed priorities of the two objectives for the second scenario, where the model selected a portfolio with lower negative deviations from the

Table 9 Selected projects for each scenario

Project	SC1	SC2	SC3	SC4
1	1	1	1	
2	1	1	1	1
3	1			1
4	1	1	1	1
5	1	1	1	1
6	1	1	1	1
7		1	1	1
8	1	1	1	1
9			1	
10	1	1	1	1
11				1
12	1	1	1	1
13	1	1	1	1
14	1	1	1	1
15				1
Total	**11**	**11**	**12**	**13**
Objective (%)	**35.6**	**42.3**	**12.6**	**0.0**

where SC1 to SC4 presents the projects selected in scenarios 1 to 4

proposed targets, using fewer resources in relation to the previous situation, though it was still within budget. Aiming at a greater reduction of deviations, in the third scenario, the positive deviation of the budget target began to have a lower priority in relation to other priorities. This scenario resulted in the lowest total negative deviation of the portfolio but with a budget overflow. Finally, in the last scenario, the question of how much more budget would be needed for non-negative deviation in other goals was evaluated.

The comparison among the four situations allows for evaluating and choosing the best selection of proposed projects. The trade-offs performed among each of the four situations assist in decision making since the decision maker can evaluate the reduction of deviation of a goal to the detriment of the increase or reduction of the deviations of others. Clearly, there is no better situation among the four presented, and the choice of best is in the hands of the decider based on his/her needs and availabilities. The relationships of projects selected by the model in the four evaluated situations is presented in Table 9, indicating the total of the selected projects and the value obtained for the objective function.

Figure 3 shows deviations of the goals of each objective in the 4 scenarios. To facilitate the presentation of these deviations, it was agreed that positive deviations of the goals would be expressed above the figure axis and negative deviations of

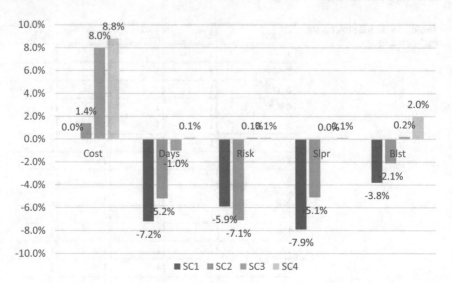

Fig. 3 Deviation goals per objective and scenario. Where O = Budget, NPV = Net Present Value, DF = Physical Availability, AC = Accident Rate Decrease and U = Attendance to stakeholders

targets would be represented below the figure axis. However, all the deviations have positive values.

3.2 Railway Superstructure Maintenance Selection

Maintenance activity is an important part of the total cost of railway business. In this context, maintenance of the railway system can be considered a strategic activity inherent in the transportation of cargoes since results are directly related to performance, reliability, transport safety and cost reduction. Problems related to the prioritization of activities are usually part of the railway maintenance process and involve goals and constraints such as budget constraints, the availability of labour and resources, and the degradation of permanent track materials.

Railway maintenance interventions are essentially divided into three unique operations: preventive maintenance, one-off interventions and track renewals. In the first situation, deteriorated track components are checked, followed by repairs and/or replacement. However, punctual interventions are corrective maintenance related to some fault of a track component that compromises train circulation, such as rail fracture fixes, the repair of geometric deformations, and the replacement of screws. Finally, railway renovation is a process in which all elements are replaced due to the inability of the railway to function or to increase transport capacity, considering traffic conditions that are higher than those existing.

In general, railway maintenance is a process of organizing the maintenance of a permanent track to keep railway superstructure in good operating condition. According to (Ferreira 2010), with robustness and high investment involved in the maintenance process, railroads started to adopt maintenance investment strategies increasingly directed to critical problems that were identified. However, a misconception in the railway maintenance process can result in defects concentrated in weak points that, if not properly identified and repaired, can create permanent and irreversible deformations in the railway. The solution demands the complete replacement of components of a whole stretch with possible unwanted interruptions in the movement of trains. The choice of location that requires maintenance intervention is usually performed due to the degradation of permanent track components that occur mainly due to fatigue and the wear actions of material together with the speed of train circulation in the stretch, type of component material, track geometry, and geographic and climatic conditions, etc.

3.2.1 Definition of Model Objectives

The definition of objectives is based on a general objective, which, in turn, is broken down into specific objectives until each can be measured by some attribute. Therefore, the proposed general objective (first level) of this application is "To prioritize maintenance of railway superstructure according to available resources, workforce, risks and maintenance indicators". The general objective can be broken down into the following specific objectives in the second level (Table 10).

Objective 1.3 does not have a unit that describes the risk associated with the region or location where the defect is due to be a qualitative objective, and it is necessary to use the Associated Utility Method for a better description in terms of a numerical scale. Thus, the following environmental, demographic and geometric factors are considered:

- Proximity to water courses or environmental preservation areas;
- Proximity to urban centres or towns;
- Railway Geometry (straight or curved).

Table 10 Specific objectives of the second level

ID	Objective	Attribute	Unit
1.1	Do not exceed limit of financial resources available for maintenance of railway	Resource	R$
1.2	Do not exceed work capacity available for maintenance activities	Labour	days
1.3	To reduce risks of road defects associated with defect localization	Risk index	ur
1.4	To improve current track condition according to maintenance indicators	–	–

Table 11 Specific objectives of the third level

ID	Objective	Attribute	Unit
1.4.1	To replace insufferable sleepers	Sleeper	un
1.4.2	To replace rails below tolerance levels	Rail	m
1.4.3	To apply ballast to sections with insufficient materials	Ballast	m

The estimation of risk index involves the relationship between measurement and criticality of defects (in sleeper, ballast and rail) and severity represented by location of these defects. The equation for estimation of risk index, measured in units of risk, is represented as follows:

$$R_i = \sum_{m=1}^{M}(u_m \cdot \delta_{im}) \sum_{n=1}^{N}(g_n \cdot \lambda_{in}) \tag{17}$$

where

i   each section or division of railway in I equal parts (i = 1, 2, 3,..., I);
m   type of defect analysed: sleeper (m = 1), ballast (m = 2), rail (m = 3);
n   criterion referring to defect locality: environmental, geometric (n = 1, 2, 3,..., N);
u_m criticality factor or urgency factor of defect m;
g_n weight for each of *n* evaluation criteria of locality;
δ_{im} measure or extent of defect m in section *i*;
λ_{in} factor of severity of locality referring to criterion n in section *i*.

The value of the severity factor associated with locality λ_{in} is represented by the numerical scale reported from 1 to 5.

Objective 1.4, described in a generic way in Table 10, needs more detail so that it can have a form of measurement. Thus, this item is broken down into more third level specific objectives (Table 11), as follows:

3.2.2   Transformation of Objectives into Goals

Once objectives are defined, they must be transformed into goals. The goal is achieved by including negative and positive deviations in mathematical expressions that represent the goal and also by attribution of its target or level of attainment. For the problem in question, the following goals are formulated:

$$\text{Cost} \quad \sum_{i=1}^{I}(x_i \cdot C_i) + \eta_1 - \rho_1 = M_1 \tag{18}$$

$$\text{Labour} \quad \sum_{i=1}^{I}(x_i \cdot W_i) + \eta_2 - \rho_2 = M_2 \tag{19}$$

$$\text{Risk reduction} \quad \sum_{i=1}^{I}(x_i \cdot R_i) + \eta_3 - \rho_3 = M_3 \tag{20}$$

$$\text{Sleeper replacement} \quad \sum_{i=1}^{I}(x_i \cdot S_i) + \eta_4 - \rho_4 = M_4 \tag{21}$$

$$\text{Rail replacement} \quad \sum_{i=1}^{I}(x_i \cdot T_i) + \eta_5 - \rho_5 = M_5 \tag{22}$$

$$\text{Ballast application} \quad \sum_{i=1}^{I}(x_i \cdot B_i) + \eta_6 - \rho_6 = M_6 \tag{23}$$

where

i	each section or division of railway in I equal parts ($i = 1, 2, 3,\ldots, I$);
j	numerical identification of each goal ($j = 1, 2, 3,\ldots, J$);
η_j and ρ_j	variables of negative and positive deviation of goal j;
M_j	target values for each goal j;
x_i	binary variable of decision for each section i;
C_i	investment or cost required to perform maintenance in each section i [R\$];
W_i	labour or workforce required to maintain each section i [days];
R_i	risk index associated with defects in each section i [ur];
S_i	number of sleepers to be applied in each section i [und];
T_i	extension of rails to be replaced in each section i [m];
B_i	extension of ballast to be replaced in each section i [m].

3.2.3 Model Formulation

Once goals are defined, an objective function is proposed that will seek to minimize the weighted sum of the percentage deviation variables of the target values defined for each of the goals. In this way, the decision maker can prioritize goals, normalized in percentage, which will be weighted so that deviations are smaller for most important goals. The percentage treatment of deviation weights becomes important because it allows for a comparison of deviations at the same level of magnitude. Therefore, according to Rehman and Romero (1984), the execution function and its goals are defined as follows:

$$\min \sum_{j=1}^{6}\left(\alpha_j \frac{\eta_j}{M_j} + \beta_j \frac{\rho_j}{M_j}\right) \tag{24}$$

Subject to:

$$\sum_{i=1}^{I} (x_i \cdot C_i) + \eta_1 - \rho_1 = M_1$$

$$\sum_{i=1}^{I} (x_i \cdot W_i) + \eta_2 - \rho_2 = M_2$$

$$\sum_{i=1}^{I} (x_i \cdot R_i) + \eta_3 - \rho_3 = M_3$$

$$\sum_{i=1}^{I} (x_i \cdot S_i) + \eta_4 - \rho_4 = M_4$$

$$\sum_{i=1}^{I} (x_i \cdot T_i) + \eta_5 - \rho_5 = M_5$$

$$\sum_{i=1}^{I} (x_i \cdot B_i) + \eta_6 - \rho_6 = M_6$$

where

$\eta_j, \rho_j \geq 0$ for all j
$M_j > 0$ for all j
$0 < \alpha_j \leq 10$
$0 < \beta_j \leq 10$
$x_i = 0$ or $x_i = 1$
where α_j and β_j weights are negative and positive deviations of goal j

The proposed model was implemented in Microsoft Excel solver due to the ease in formulating and changing data for generations of different scenarios and because it is a commercially distributed software.

3.2.4 Model Application

The proposed model for prioritization of railway sections that will undergo maintenance intervention is then applied to a 100-km-long fictitious railroad, which is divided into 100 stretches of 1 km each, and the number of defects of randomly simulated superstructure materials with values is presented in Table 12.

The GP model considers the division of the railway into equal parts and seeks to indicate, from decision variables, which sections should have maintenance prioritized to minimize percentage deviations of target values for each goal. Each division has a necessary amount of material, labour (in working days) and investment, as well as respective risk associated with the need for maintenance and factors inherent in location of defects.

Table 12 Simulated railroad parameters

Item	Amount
Total extension (km)	100
Sequences of insufferable sleepers (unid)	60,885
Extension of rails below tolerance levels (m)	28,566
Extension of insufficient ballast (m)	36,995

Table 13 Proposed model targets

	Sections (unit)	Investment (BzR$)	Work Days	Risk Reduction	Sleepers (unit)	Ballast (m)	Rails (m)
Required	100	27,257,332	555	681,150	60,885	36,995	28,566
Target	–	20,442,999	472	578,978	51,753	31,446	24,281
Target (%)		75	85	85	85	85	85

As in practical situations, there are budget constraints. It is considered a strategy of the decision maker that the available budget for selection of stretches to be worked is approximately 75% of the real need and that target values of other goals are equal to 85% of the value total of each attribute (Table 13); it is the responsibility of the decision maker to carry out trade-offs to evaluate possible extrapolations of the available budget.

To analyse different trade-offs, from possible changes in priorities of objectives carried out by the decision maker, different scenarios are generated to be analysed. For the First Scenario, according to the strategic planning for railroad maintenance, objectives contained in the project selection model have the following priority levels:

- Priority 1: Investment;
- Priority 2: Working days;
- Priority 3: Risk reduction;
- Priority 4: Sleepers application;
- Priority 5: Ballast application;
- Priority 6: Replacing rails.

In this way, the model will select the segment that will receive investment intervention considering that the greater the value of the weights of deviation variables of a goal, the higher the level of its priority because the objective function of the model will seek to minimize the weighted sum of percentage deviation variables of target values.

Using these assumptions, for the First Scenario, the model selected 76 stretch from the existing 100, which are presented in Table 14. In this case, the only goal that did not present a deviation was one of investment, considering that its overflow would be acceptable and receive maximum weight. All other goals had a deviation below their target values.

Table 14 First maintenance scenario results

Objective	Required	Target	Amount reached	Difference	+Deviation (%)	−Deviation (%)	Weight above	Weight below
COST	12,243,418	20,442,999	20,444,856	1857	0.0	0.0	10	0
DAYS	276	472	438	−34	0.0	7.2	0	1
RISK	284,067	578,978	544,744	−34,234	0.0	5.9	0	5
SLPR	33,000	51,753	47,675	−4078	0.0	7.9	0	3
BLST	16,800	31,446	30,252	−1194	0.0	3.8	0	3
RAIL	11,000	24,281	20,470	−3811	0.0	15.7	0	3

In a second scenario, with inversion of orders of priority of material application and risk reduction objectives, while maintaining restrictions to exceed available investment, it is possible to note that the solution found is a 1% overflow in the budget that generated a lower reduction in existing risk and greater application of materials. From this change, the model selected 77 stretches compared to 76 from the first scenario (Table 15).

The third scenario (Table 16) prioritized the application of materials and risk reduction, making the investment goal more flexible. In this way, the reduction of deviations from this goal was quite pronounced, resulting in an 8% overflow in budget and prioritizing maintenance in 82 sections.

The fourth scenario (Table 17), considered as a priority optimization of available labour, increases the weight of working days and reduces investment priority and risk, maintaining some importance of the application of materials. Thus, the model selected prioritization of maintenance in 83 stretches, zeroing deviations of the goal of days worked and presenting applications of sleepers with deviations close to zero and applications of ballast superior to that established as the goal. The number of rails to be replaced presented a deviation of less than 8%, which can be explained by its high unit value.

3.2.5   Results Analysis

The first scenario (Table 14) was the object of mainly available budget consultation and, from results achieved, was reversed as direct from other objectives for the generation of a second scenario (Table 15), a variation of 1.4% in positive deviation of target budget. Most of the deviations are not found in the third scenario (Table 16); a positive deviation from budget target became less important in relation to others. There is nothing less than this negative in total portfolio, but without budget. Finally, the maximum available labour force was evaluated in the last scenario (Table 17).

A comparison between four series allows us to evaluate and choose a better selection of proposed maintenance strategies. Trade-offs in one of four situations help in decision making since the decision maker can evaluate the reduction of the deviation of one goal in detriment of high or the reduction of deviations of others. It is important that its opinions be more rigorous, improved and assessed for priority.

Finally, Fig. 4 shows the deviation goals of each of the 4 scenarios. The most important of these are for status of representative of axis. However, all deviations have positive values.

Table 15 Second maintenance scenario results

Objective	Required	Target	Amount reached	Difference	+Deviation (%)	−Deviation (%)	Weight above	Weight below
COST	12,243,418	20,442,999	20,729,629	286,630	1.4	0.0	10	0
DAYS	276	472	447	−25	0.0	5.2	0	1
RISK	284,067	578,978	537,862	−41,116	0.0	7.1	0	2
SLPR	33,000	51,753	49,108	−2645	0.0	5.1	0	5
BLST	16,800	31,446	30,777	−669	0.0	2.1	0	5
RAIL	11,000	24,281	20,496	−3785	0.0	15.6	0	5

Table 16 Third maintenance scenario results

Objective	Required	Target	Amount reached	Difference	+Deviation (%)	−Deviation (%)	Weight above	Weight below
COST	12,243,418	20,442,999	22,087,474	1,644,475	8.8	0.0	2	0
DAYS	276	472	467	−5	0.0	1	0	1
RISK	284,067	578,978	579,558	580	0.1	0.0	0	5
SLPR	33,000	51,753	51,742	−10	0.0	0.0	0	5
BLST	16,800	31,446	31,495	49	0.2	0.0	0	5
RAIL	11,000	24,281	22,188	−2093	0.0	8.6	0	5

Table 17 Fourth maintenance scenario results

Objective	Required	Target	Amount reached	Difference	+Deviation (%)	−Deviation (%)	Weight above	Weight below
COST	12,243,418	20,442,999	22,234,093	1,791,094	8.8	0.0	1	0
DAYS	276	472	472	0	0.1	0.0	0	5
RISK	284,067	578,978	579,625	648	0.1	0.0	0	1
SLPR	33,000	51,753	51,787	34	0.1	0.0	0	3
BLST	16,800	31,446	32,090	644	2.0	0.0	0	3
RAIL	11,000	24,281	22,382	−1899	0.0	7.8	0	3

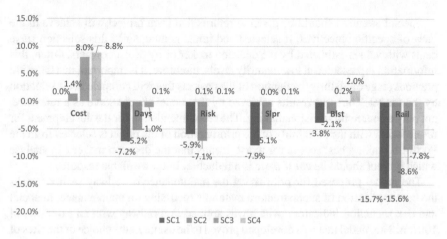

Fig. 4 Percentage deviation goals in each scenario

4 Conclusions

The use of optimization techniques by multiple criteria has grown every year in various fields of application. Among these techniques, programming by goals (GP) stands out. In the GP approach, goals are formulated by the association of targets to be achieved in each objective, having an objective function that seeks to minimize the sum of absolute deviations of these goals.

Based on the GP methodology, a model of the selection of current investment projects was developed, seeking the composition of a project portfolio focussing on strategic-level financial and sustainability indicators, which was applied to the selection of a portfolio of 15 fictitious projects. To evaluate the trade-offs carried out with a change in the priority of the objectives, 4 selection scenarios were generated from the same project portfolio, varying the weights of the variable deviations of the targets, which directly influences programming results. The scenarios were generated in sequence, from an analysis of the variables of the deviation of the targets and not the value obtained by the objective function of each previous scenario since each change of the weights of the deviation variables in each scenario results in different situations.

The developed model proved useful in the selection of railway investment projects, but it is only a tool to assist the decision maker, who needs to define which goals need to be prioritized according to business needs. The target values of each goal must be defined according to the strategy of the organization, and the change of these values directly influences the results achieved by the model. In addition to prioritizing the most important goals, based on model results, the decision maker can also redefine the target values of their goals, seeking harmony with the possible results against existing objectives.

Special attention should be given to returns that each project will achieve in the view of specific objectives, if selected and implemented, since this selection stage deals with values estimated by the decision maker or by project owners. Often, this information is presented at low maturity levels; therefore, it is important that there is a previous stage of maturity evaluation of the projects that will compete in the selection by the model. Thus, the results achieved by the model should serve as a reference, not only a source for decision making. The recommendation for the development for future work, with the possibility of the prioritization of the projects selected from the Portfolio ranking best scored for worst, means that the decision maker can analyse which project should be cut if there is a reduction in the available resources.

The model proposed for prioritizing the maintenance of railway sections seeks the best application of superstructure materials focussing on maintenance, financial and risk reduction indicators, which was applied in a railway with an extension of 100 km. The model that was developed proved to be useful in the choice of the sites of the application of the superstructure materials as a tool to assist the decision maker.

For these two applications, the decision maker must define the prioritization of the goals according to the strategy of the organization. The change of these values directly influences the results achieved by the models.

The software used proved to be very friendly and suitable for implementation of the model. However, the software can present limitations in more robust applications due to the complexity of the problem to be modelled, such as additions of variables or goals, as well as the greater number of subdivisions of model.

The models developed were used to select projects for railway investments and maintenance strategies, but they could be applied in other applications if the specific objectives and target values of the goals are duly redefined.

Acknowledgements Part of these studies were funded by FAPES N° 04/2015

References

Ahern A, Anandarajah G (2007) Railway projects prioritisation for investment: application of goal programming. Transp Policy J 14:70–80. https://doi.org/10.1016/j.tranpol.2006.10.003

Ahern A, Anandarajah G (2008) An optimisation model for prioritising transport projects. Proc Inst Civ Eng – Transp 161:221–230. https://doi.org/10.1680/tran.2008.161.4.221

Aouni B, Colapinto C, La Torre D (2014) Financial portfolio management through the goal programming model: current state-of-the-art. Eur J Oper Res 234:536–545. https://doi.org/10.1016/j.ejor.2013.09.040

Chang Y, Wey W, Tseng H (2009) Using ANP priorities with goal programming for revitalization strategies in historic transport: a case study of the Alishan Forest Railway. Expert Syst Appl 36(4):8682–8690. https://doi.org/10.1016/j.eswa.2008.10.024

Charnes A, Cooper WW (1961) Management models and industrial applications of linear programming. Willey, New York

Charnes A, Cooper WW, Ferguson RO (1955) Optimal estimation of executive compensation by linear programming. Manage Sci, 138–151

Colapinto C, Jayaraman R, Marsiglio S (2015) Multi-criteria decision analysis with goal programming in engineering, management and social sciences: a state-of-the art review. Ann Oper Res 251:7–40. https://doi.org/10.1007/s10479-015-1829-1

Ferreira JA (2010) Intervenções de construção, renovação e manutenção na via-férra. Dissertação de mestrado – Programa de Pós-Graduação em Engenharia Civil da Faculdade de Engenharia da Universidade do Porto, Porto

Jones D, Tamiz M (2010) Practical goal programming, 1ª edn. Springer, New York

Kahraman C, Büyüközkan G (2008) A combined fuzzy AHP and fuzzy goal programming approach for effective six-sigma project selection. J Multiple-valued Logic Soft Comput 14:599–615

López-Ramos F, Codina E, Marín A, Guarnaschelli A (2017) Integrated approach to network design and frequency setting problem in railway rapid transit systems. Comput Oper Res 80:128–146. https://doi.org/10.1016/j.cor.2016.12.006

Margueron M (2003) Processo de tomada de decisão sob incerteza em investimentos internacionais na exploração & produção de petróleo: uma abordagem multicritério. Dissertação de mestrado – Curso de Pós-Graduação em Engenharia da Universidade Federal do Rio de Janeiro, Rio de Janeiro

Morais Neto G (1988) Sistema Decisório Interativo de Alocação de Fluxo de Cargas. Dissertação de mestrado – Curso de Pós-Graduação em Sistemas e Computação do Instituto Militar de Engenharia, Rio de Janeiro

Niemeier D, Zabinsky Z, Zeng Z, Rutherford G (1995) Optimization models for transportation project programming process. J Transp Eng 121:14–26. https://doi.org/10.1061/(ASCE)0733-947X(1995)121:1(14)

Ramos A (1995) Procedimento para tomada de decisão em terminais marítmos petroleiros. Dissertação de mestrado – Curso de Pós-Graduação em Sistemas e Computação do Instituto Militar de Engenharia, Rio de Janeiro

Rehman T, Romero C (1984) Multiple-criteria decision-making techniques and their role in livestock ration formulation. Agric Syst 15(1):23–49. https://doi.org/10.1016/0308-521X(84)90016-7

Tamiz M, Jones D, El-Darzi E (1995) A review of goal programming and its applications. Ann Oper Res. Springer, pp 39–53. https://doi.org/10.1007/BF02032309

Tamiz M, Jones D (1998) Goal programming for decision making: an overview of the current state-of-the-art. Eur J Oper Res 111:569–581. https://doi.org/10.1016/s0377-2217(97)00317-2

Teng J, Tzeng G (1998) Transportation investment project selection using fuzzy multiobjective programming. Fuzzy Sets Syst 96:259–280. https://doi.org/10.1016/S0165-0114(96)00330-2

Teng J-Y, Tzeng G-H (1996) A multiobjective programming approach for selecting non-independent transportation investment alternatives. Transp Res Part B: Methodol 30(4):291–307. https://doi.org/10.1016/0191-2615(95)00032-1

Uliana A (2010) Utilização de Programação por Metas como Auxílio à Tomada de Decisão na Distribuição de Gás Natural. Dissertação de mestrado – Programa de Pós-Graduação em Engenharia Civil da Universidade Federal do Espírito Santo, Vitória

Valle A, Soares C, Finocchio J, Silva L (2007) Fundamentos do Gerenciamento de Projetos, 1ª edn. FGV Editora, Rio de Janeiro

Vargas R (2010) Using the analytic hierarchy process (AHP) to select and prioritize projects in a portfolio. In: PMI Global Congress. Project Management Institute, Washington, pp 1–22

Wey W, Wu K (2007) Using ANP priorities with goal programming in resource allocation in transportation. Math Comput Model 46:985–1000. https://doi.org/10.1016/j.mcm.2007.03.017

Yang X, Low J, Tang L (2011) Analysis of intermodal freight from China to Indian Ocean: a goal programming approach. J Transp Geogr 19:515–557. https://doi.org/10.1016/j.jtrangeo.2010.05.007

Parallel Genetic Algorithm and High Performance Computing to Solve the Intercity Railway Alignment Optimization Problem

Cassiano A. Isler and João A. Widmer

Abstract Despite the advances in solving the Railway Alignment Optimization (RAO) problem, the computational burden of the current algorithms to estimate new intercity connections of minimum cost is still an issue. This paper proposes a parallel Genetic Algorithm framework running on a high performance computing environment to solve the RAO problem while minimizing the costs of new railway alignments constrained by the geometric parameters required to run trains with different average speeds. The framework was applied to new connections between Brazilian cities and the results show that it is capable of providing accurate estimations compared to the international experience. From the computational aspect, the parallel computing approach drastically reduces the running times in the cases studied. However, scaling the computing infrastructure to more than 5 machines running in parallel may not be advantageous since the running times do not decrease significantly when more virtual machines are available.

Keywords Railway · Alignment · Parallel Genetic Algorithm · High Performance Computing

1 Introduction

Railway transport is capable of influencing economic and social growth of regions, as passenger transport may lead to the development and growth of different activities such tourism, culture etc., and freight transport can reduce road usage (Dolinayova et al. 2018). However, planning new railway alignments (i.e., the sequence of straight lines connected by curves) is complex and usually comprises an iterative process of

C. A. Isler (✉)
Transportation Engineering Department, Polytechnic School, University of São Paulo, Av. Prof. Almeida Prado, 83, Travessa 2, São Paulo, SP, Brazil
e-mail: cassiano.isler@usp.br

J. A. Widmer
Transportation Engineering Department, São Carlos School of Engineering, University of São Paulo, Av. Trabalhador São-Carlense, 400, São Carlos, SP, Brazil

© Springer Nature Switzerland AG 2020   159
M. Marinov and J. Piip (eds.), *Sustainable Rail Transport*, Lecture Notes in Mobility,
https://doi.org/10.1007/978-3-030-19519-9_5

a multidisciplinary team of specialists and, thus, is not straightforward and hardly leads to near-optimal costs without computational assistance.

According to Li et al. (2016), the railway alignment optimization (RAO) problem is defined as the task of finding the sequence of horizontal and vertical curves connected by straight lines while minimizing a mathematical function of total cost given the geometric and operational constraints of the transportation mode.

OECD (1973) and Chew et al. (1989) classify the costs to build highways or railway alignments into infrastructure, earthwork, tunnels and bridges, expropriation and drainage; where the first three account for 75% of the overall construction costs. Additionally, Schonfeld et al. (2007) separate the infrastructure cost into building and operational costs to users and to the operator.

It is well known that an appropriate alignment optimization model must compute the most significant cost items such as earthwork and infrastructure (tunnels and bridges). It should not violate the geometric constraints related to the mode of transport and its rolling stock. It also should simultaneously optimize 3-dimensional alignments, search within a continuous solution space, yield a realistic alignment, have an efficient solution algorithm in terms of memory requirements and computing time, must be compatible with a Geographic Information System (GIS) and should avoid inaccessible regions (Jha et al. 2006).

Besides the geological, hydrological, land use and topographical conditions provided by georeferenced databases, the safety and riding comfort standards of trains are relevant aspects to be considered when optimizing railway alignments. However, given these georeferenced datasets, the computational burden to solve the RAO problem requires significant efforts to find good quality solutions.

In this paper, a parallel Genetic Algorithm running on a high performance computing environment is proposed to solve the RAO problem while minimizing the costs of new railway alignments constrained by the geometric parameters required to run trains with different average speeds on intercity connections. The framework was applied to different connections between Brazilian cities considering High Performance Trains (henceforth HPTs) with average speeds of 200 km/h and High Speed Trains (onwards HSTs) running at 300 km/h on average. This application was undertaken in order to assess its accuracy in estimating the costs of intercity railway alignments and in evaluating its benefits in terms of processing times to achieve good solutions.

While the approach is based on previous work, it is not merely an adaptation of the existing solutions to the RAO problem since: (i) a parallel approach over a set of high performance computers is proposed to estimate the costs of new railway alignments; (ii) different geometric constraints and types of trains distinguished by their average speeds are taken into account to estimate the alignments; and (iii) the framework is applied to estimate the costs of intercity alignments in different topographical and land use conditions in Brazil.

The remainder of this paper is organized as follows. Section 2 presents a literature review of previous research on RAO and solution strategies to solve the problem. Section 3 describes a Genetic Algorithm (GA) to solve the model, and the proposed parallel computing framework, followed by Sect. 4 where the results of its application to different connections in Brazil are shown and compared with values from the literature based on international practice. Finally, Sect. 5 summarizes the conclusions of the research.

2 Literature Review

The RAO is distinguished from the highway alignment optimization problem by the objective function to be minimized. However, both mathematical models address the horizontal and vertical curves as continuous differentiable functions, constrained by minimum and maximum radii and slopes at their respective derivatives in successive points of the alignment (Jha et al. 2006).

The mathematical models and solution methods to these alignment optimization problems emerged in the 1970s and were applied to different contexts in order to assist the decision makers planning new transport infrastructures. However, the solution to these problems can hardly be achieved to optimality by an exact method since the terrain configuration usually cannot be represented as a continuous surface. Moreover, the design parameters regarding the minimum values of the horizontal and vertical alignments also affect these estimations (Hodas 2014).

Several mathematical models and heuristic algorithms were proposed to solve the horizontal, vertical, and three-dimensional alignment optimization problems (Li et al. 2016, 2017): calculus of variations (Howard et al. 1968); enumeration (Easa 1988); dynamic programming (Li et al. 2013); genetic algorithms (Jha 2003; Kang et al. 2012); neighborhood search heuristics with mixed integer programming (Cheng and Lee 2006); mixed integer programming (Easa and Mehmood 2008); particle swarm optimization (Maji 2017); and distance transform (De Smith 2006; Li et al. 2016, 2017).

The benefits and shortcomings of these methods are addressed by Jha et al. (2006) to optimize highway alignments, which may be extended to the RAO problem. However, despite being the most promising approach to solve the problem in reasonable computational time, Genetic Algorithms (GA) still require large computational resources to assess the large-scale datasets containing information on the topographic and land use conditions.

Jha and Schonfeld (2000) used a Geographic Information System (GIS) to assess the land use costs to build a new highway infrastructure, and Jha et al. (2001) and Jha (2003) proposed a decision support system that enables the alteration of an alignment given the surrounding infrastructure. Jha et al. (2007) applied the GA to the RAO using the method proposed by Jha et al. (2006), where the objective function comprised the operator costs (track construction, stations, earthwork, land use and operational costs) and the costs to users (access, egress, and travel time).

Li et al. (2016) proposed a two-phase methodology to solve the RAO problem, where promising paths are generated in the first phase, followed by curve refinements. The approach is validated through a real-world case study in a mountainous area where the natural terrain gradient is nearly the triple of the maximum allowed design gradient.

Samanta and Jha (2011) proposed a model to plan a rail transit line in which the optimal alignment is obtained by microscopic analysis, followed by the solution of a station location problem by a Genetic Algorithm which minimizes the total system cost per person, the total user cost per person and maximizes the total ridership.

Lai and Schonfeld (2016) and Pu et al. (2018) applied a distance-transform algorithm to solve the railway alignment and station location problem concurrently in the urban context and in mountainous terrain, respectively. The concurrency relates to the simultaneous alignment optimization and station location.

Kim et al. (2005) considered dividing a studied area into smaller regions to deal with the computational burden to solve the highway optimization problem through the so-called "Stepwise Genetic Algorithm". However, the evidence through statistical hypothesis testing proved its effectiveness only to a small theoretical grid area of 200 × 200 ft.

The parallel computing approach has been applied to solve combinatorial optimization problems with different methods such the Variable Neighborhood Search and the Bee Colony Algorithm (Gupta and Deep 2009; Crainic et al. 2012). The adaptive Genetic Algorithm based on a multi-population parallel approach proposed by Chen et al. (2011) is capable of estimating the costs of highway alignments while avoiding premature convergence compared to a single-population evolutionary algorithm. Kazemi and Shafahi (2013) solved the highway alignment optimization problem with a parallel processing particle swarm optimization algorithm.

As far as we know, the proposed approach to solve the alignment optimization problem through a parallel Genetic Algorithm running on a set of high performance computers has not been explored in the literature, considering the computational burden to estimate new railway alignments in wide areas, such intercity connections in large countries as Brazil. Nowadays, data can be easily stored and processed by interconnected computers in data warehouses, such the parallel computing consists of a physical infrastructure of computers (Virtual Machines, VMs) remotely accessed and an interface that enables exchanging information among servers and clients through the Internet.

3 The Parallel Genetic Algorithm and High Performance Computing Environment

This section details the parallel computing framework considered to obtain near-optimal railway alignments. The literature review presented in last section shows that the Genetic Algorithm has been extensively applied to solve both the highway and railway alignment optimization problems. One of the main contributions in this field is by Jha et al. (2006), who detailed the mathematical formulation and procedures to

Set parameters
Create individuals of population (set of alignments)
Set genes of individuals (coordinates of Horizontal Intersection Points and their respective altitudes)
For each individual execute the following *SUB-ROUTINE*
    Eliminate horizontal circular curves
    Set radii of horizontal circular curves
    Calculate attributes of horizontal circular curves
    Set Vertical Intersection Points
    Calculate length of vertical parabolic curves
    Calculate attributes of vertical parabolic curves
    Calculate the position and elevation of equally spaced track points
    Calculate fitness (total cost)
Sort population by fitness (alignments by total cost)
Identify alignment of lowest cost (best individual with lowest fitness)
Calculate probability of changing or removing individual from the population
While NUMBER GENERATIONS ≠ MAXIMUM GENERATIONS do
    Calculate number of crossover and mutation operators to be executed
    Select the operators to be executed
    Crossover and mutate individuals
    For each individual execute the *SUB-ROUTINE*
    Update population
    Sort population by fitness (alignments by total cost)
    Identify alignment of lowest cost (best individual with lowest fitness)
    Calculate probability of changing or removing individual from the population
    If fitness of best individual of current iteration = fitness of best individual of last iteration **then**
    NUMBER GENERATIONS=NUMBER GENERATIONS +1
    Else
    NUMBER GENERATIONS=0
    End If
End While
Print best individual (attributes and georeferenced file of the lowest cost alignment)

Fig. 1 Steps of the genetic algorithm implemented in this paper to estimate railway alignments. *Source* Adapted from Jha et al. (2006)

estimate the overall costs to build linear transport infrastructures with applications to road design.

The steps of the 3-dimension Genetic Algorithm implemented in this paper to estimate new railway alignments are described in Fig. 1, taking into account the procedures described by Jha et al. (2006).

Initially, the algorithm sets the coordinates of the start (S) and end (E) points of the alignment, the values of track parameters (minimum horizontal radius, and minimum and maximum slope), and the GA population size (number of alignments), the number of generations (iterations of the algorithm) and the unit cost to estimate the fitness function of each individual (i.e., the cost of each alignment).

Next, the individuals of the population are created, each one representing an alignment containing Horizontal Intersection Points (HIPs) and their respective elevations defined as: points over the straight line between S and E, and random elevations; points in equidistant perpendicular plans over the straight line between S and E, and random elevations; points in equidistant perpendicular plans over the straight line between S and E, and ground elevation based on Digital Elevation Model (DEM); random points and ground elevations based on DEM; or random points and random elevations.

For each individual, the algorithm executes a *SUB-ROUTINE* where excessive horizontal curves are eliminated, the radii of the horizontal circular curves constrained by minimum values as a function of the railway technology to be operated are set, and the attributes of the tangent and curvature points of each horizontal circular curve are calculated.

The Vertical Intersection Points (VIPs) are defined in the same position of the HIPs in a way that three geometric elements may arise in these locations given the slope constrained by minimum and maximum parameters: (i) a horizontal circular curve and a vertical parabolic curve result in a three-dimension curvature; (ii) a horizontal curve stands in a flat or sloping terrain; and (iii) or a tangent section lies on a vertical curve.

Once the altitudes of the Vertical Intersection Points are defined based on the values of their respective HIPs, the length of the parabolic curves and their attributes are calculated, followed by the identification of the position and altitude of equally spaced track points along the three-dimensional alignment.

The fitness function of each individual, i.e., the overall construction cost of each alignment (CC), is estimated as the sum of the track related cost (TRC), the land use cost (LUC), the earthwork cost (EWC) and the costs to build tunnels and bridges (TBC). The TRC refers to the track elements (rails, sleepers, electrification etc.) and is calculated as a function of the total length of the alignment and an average unit cost per kilometer.

The land use cost (LUC) is calculated by the expropriation costs over a surrounding area of the alignment given an average unit cost depending on the land use provided by a georeferenced dataset (IBGE 2014), which classifies the studied region into urban and rural areas as illustrated in Fig. 2 (left). The cutting and embankment volumes calculate the earthwork costs (EWC) over successive track points given their cross-sectional areas and the elevations provided by the DEM illustrated in of Fig. 2 (right) (*U.S. Geological Survey* 2014).

Finally, the costs to build tunnels and bridges (TBC) are estimated based on unit monetary values per kilometer along the sequence of track points where those structures are more economical than earthworks. An economic break-even point between earthwork cost and construction cost of bridges or tunnels is determined by the difference between the ground elevation and the altitude of the equally spaced track points of the vertical alignment.

By the end of the *SUB-ROUTINE*, all the individuals are sorted in ascending order of their fitness, and the probability of changing their genes is calculated based on an uniform distribution. Four types of crossover operators (simple, two-point, arithmetic and heuristic) and four types of mutation operators (uniform, straight and non-uniform mutation to one or to all Horizontal Intersection Points) addressed by

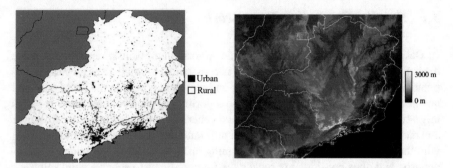

Fig. 2 Georeferenced datasets of land use (left) and DEM (right) of the Brazilian Southeastern region in Brazil

Jha et al. (2006) are randomly applied to 30–50% of the population. At each crossover or mutation, the new individuals are submitted to the *SUB-ROUTINE* to calculate their fitness.

The replacement of individuals of the old population with a new one has been implemented based on Jha et al. (2006). A random value between zero (0) and one (1) is drawn based on an uniform probability function and assigned to the new individual obtained by the crossover or mutation. This value is then compared with a calculated probability of excluding the kth individual of the existing population based on Eq. (1). If the assigned random value is between p_k and p_{k+1}, then the old individual in the kth position of the old population sorted in the descending order of fitness is replaced by the new one, otherwise the new individual is excluded. This replacement procedure occurs after performing all the crossovers and mutations.

$$p_k = \frac{q \cdot (1-q)^{k-1}}{1 - (1-q)^{n_p}} \tag{1}$$

where p_k = choice probability of the kth individual sorted in descending order of the fitness function; n_p = population size; and q = exchange parameter equals 0.25 as recommended by Jha et al. (2006).

The new population is sorted once again in ascending order of fitness and the first individual is identified, which represents the alignment of lowest total cost. Finally, a stop criterion is checked: if the number of successive iterations in which the value of the fitness function of the best individual among the population does not change, then the parameter *NUMBER_GENERATIONS* is increased by one unit, else its value is set to zero. In the former case, the crossover and mutation probabilities are re-calculated, and these operators and the *SUB-ROUTINE* are executed until the parameter *NUMBER_GENERATIONS* differs from *MAXIMUM_GENERATIONS*.

3.1 Parallel Computing Framework

Despite the effectiveness of the Genetic Algorithm previously described to solve the RAO problem, its application usually is constrained to small areas given the computational burden due to data processing and recurrent access to the DEM and land use georeferenced files. Additionally, the algorithm applied to wide areas requires large-scale datasets to be processed at each generation after executing the crossover and mutation operators. Thus, the parallel programming is a suitable approach to deal with these computational issues by assigning tasks to multiple computers simultaneously, and, thus considerably reduce the running times to achieve good solutions to the problem.

The computational experiments presented in this paper were performed using the high performance computing resources of the University of São Paulo's Advanced Scientific Computing Laboratory (LCCA). More specifically, a cluster with physical servers (virtual machines) Intel(R) Xeon(R) CPU E7-2870 @ 2.40 GHz 32 GB of RAM. The georeferenced DEM and land use datasets were stored in one of these machines and accessed by a relational programming language (MySQL 2014). The fitness function of the individuals were assessed simultaneously in different virtual machines, provided their availability previously defined by the user.

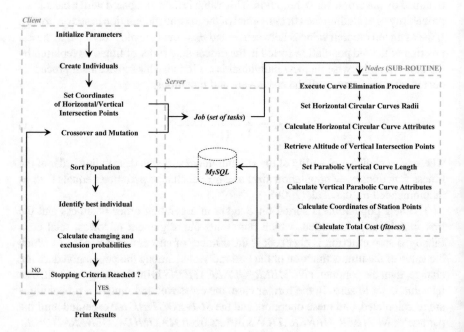

Fig. 3 Parallel genetic algorithm framework to solve the RAO problem

A virtual machine (master) containing the core of the GA coded in Java is connected to several machines (nodes) through the "Java Parallel Programming Framework" library (JPPF 2014), each one running a *SUB-ROUTINE* (task) to a specific individual.

The flowchart of Fig. 3 represents the communication among the master and the nodes of the high performance computing environment to illustrate the proposed parallel Genetic Algorithm framework to solve the RAO problem.

4 Model Application

This section presents the results of the proposed parallel GA applied to the high performance computing environment in different railway connections between Brazilian cities. The algorithm was applied to three pairs of cities in the Southeastern Region of the country, varying the number of Horizontal Intersection Points as a function of the length between them and the number of virtual machines (nodes) available to run the *SUB-ROUTINE*. The total cost, processing time, length of the alignments and the average cost per kilometer have been assessed between each city given the specified type of train (HPT and HST).

Figure 4 illustrates the location of cities in the Brazilian Southeastern Region chosen to be connected by new railway alignments, defined as "Rio de Janeiro-Juiz de Fora", "Campinas-Poços de Caldas" and "Araraquara-Ribeirão Preto". Since these cities already have a railway infrastructure in the urban areas as a consequence of their historical development, the new alignments resulting from the parallel GA were estimated only in the rural areas between them.

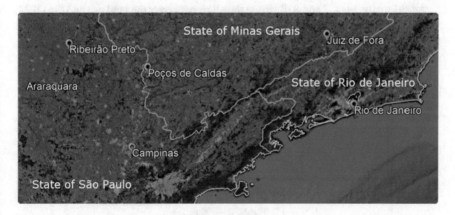

Fig. 4 Cities among which new alignments have been estimated

4.1 Parameters

For each railway technology (HPT and HST) and each pair of city, the number of Horizontal Intersection Points have been set proportionally to an average density of points per kilometer (i.e., HIPs separated by 5, 10, 15, 20 or 25 km on average) given the distance of the straight line between the start and end point of the alignment. In addition, the number of VMs available in the high performance computing environment to process the tasks in every iteration of the *SUB-ROUTINE* were set to: 1 (equivalently to execute the algorithm in a single computer), 5, 10, 25, or 50 VMs.

Finally, in order to assess the variability of the total estimated costs, the GA was executed five times per city connection, type of train, average density of HIPs, and number of available VMs. Thus, 125 executions of the GA have been performed in each studied case of intercity connection. The values of the parameters to estimate the costs of the alignments are shown in Table 1 given the standard section of the railway alignment illustrated in Fig. 5 regardless of the type of train.

4.2 Case 1: Rio de Janeiro-Juiz de Fora

Rio de Janeiro is a coastal city situated in the State of Rio de Janeiro, and Juiz de Fora is located in the State of Minas Gerais in rough terrain, and are separated by 128 km over mountainous terrain. The total costs and average costs in Brazilian monetary units (R$ and R$ per km, respectively), and processing times (seconds), and total length (km) of the estimated alignments to operate HPTs are presented in Fig. 6, obtained by the application of the proposed parallel GA framework running on the high performance computing environment.

Additionally, Figs. 7 and 8 illustrate the alignment over the land use and the DEM, and its longitudinal profile, respectively, regarding the lowest total cost solution

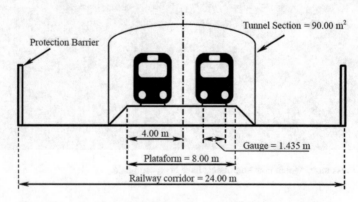

Fig. 5 Standard railway section to estimate the alignment for HPT and HST

Table 1 Parameters to estimate the railway alignments through the GA

Parameter	Genetic algorithm		Source
Non-uniformity degree (k)	6		Jha et al. (2006)
Exchange parameter (q)	0.25		Jha et al. (2006)
Population size	15. Number of HIP		Jha et al. (2006)
Stopping criterion	100 successive iterations		–
	HPT	**HST**	**Source**
Geometric			
Gauge (m)	1.435	1.435	–
Platform (m)	8	8	–
Railway corridor (m)	24	24	–
Vertical acceleration (m/s^2)	0.1829	0.1829	AREMA (2003)
Distance between track points (m)	90	90	–
Rolling coefficient	0.29	0.29	AREMA (2003)
Minimum slope (%)	−2	−2	–
Maximum slope (%)	2	2	–
Additional planning cost (%)	10	10	–
Average speed (km/h)	150	300	–
Minimum horizontal radius (m)*	2000.00	7000.00	AREMA (2003)
Earthwork			
Embankment degree (°)	45	45	TAV (2014)
Cutting degree (°)	45	45	TAV (2014)

(continued)

Table 1 (continued)

Parameter	Genetic algorithm		Source
Tunnels and bridges			
Maximum tunnel depth (m)	40	40	TAV (2014)
Maximum cutting height (m)	30	30	TAV (2014)
Maximum bridge height (m)	20	20	TAV (2014)
Maximum embankment height (m)	20	20	TAV (2014)
Minimum tunnel/bridge length (m)	50	50	TAV (2014)
Minimum tunnel/bridge length (m)	200	200	TAV (2014)
Bridge width (m)	8	8	–
Tunnel section area (m^2)	90	90	TAV (2014)
Infrastructure unit cost			
Average construction cost (R\$/m)	1556.00	4221.00	TAV (2014)
Earthwork cost			
Landfill cost (R\$/m^3)	2.71	2.71	TAV (2014)
Cutting cost (R\$/m^3)	15.4	15.4	TAV (2014)
Bulking factor	0.15	0.15	–
Landfill transportation cost (R\$/m^3 km)	1.78	1.78	TAV (2014)
Borrow pit transportation cost (R\$/m^3 km)	6.7	6.7	TAV (2014)

(continued)

Table 1 (continued)

Parameter	Genetic algorithm		Source
Tunnel/bridge cost			
Tunnel opening cost (R\$/unit)	1,660,000.00	1,660,000.00	TAV (2014)
Tunnel construction cost (R\$/m)	120,000.00	120,000.00	TAV (2014)
Tunnel construction cost (R\$/m^2)	4745.00	4745.00	TAV (2014)
Expropriation cost			
Urban area (R\$/m^2)	23.6	23.6	TAV (2014)
Rural area (R\$/m^2)	6.9	6.9	TAV (2014)

*Rounded values from 1668.0 and 6673.0 m for HPT and HST respectively, considering height difference between parallel rails equals 6 in. (approx. 0.1524 m) to compensate the centrifugal acceleration

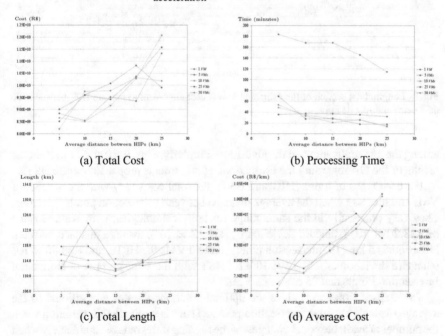

(a) Total Cost

(b) Processing Time

(c) Total Length

(d) Average Cost

Fig. 6 Average values of the estimated alignments for HPT between Rio de Janeiro and Juiz de Fora

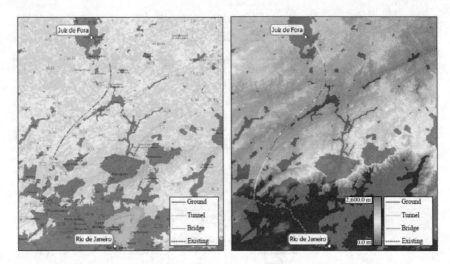

Fig. 7 Estimated lowest cost alignment over land use (left) and DEM (right) to operate HPT between Rio de Janeiro and Juiz de Fora

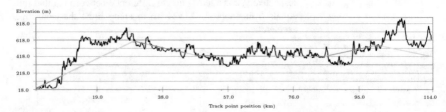

Fig. 8 Longitudinal section of the estimated lowest cost alignment to operate HPT between Rio de Janeiro and Juiz de Fora

among the 125 replications of the algorithm. Similarly, Figs. 9, 10 and 11 present the results of the GA regarding the estimations of alignments proper to operate HSTs.

For both types of trains (HPT and HST), the total costs vary with the number of VMs and increase when the average distance between intersection points increases. They vary practically in the same range for both technologies, with average value of R\$8.20 $\times$ 10^9 (standard deviation of R\$1.12 $\times$ 10^9), average distance between HIPs equals 5 km (19 intersection points) and 50 VMs for HPTs, and R\$8.50 $\times$ 10^9 (standard deviation of R\$1.12 $\times$ 10^9) for HST with the same number of Horizontal Intersection Points and 25 available VMs.

The elapsed times to achieve the optimal solutions are not proportional to the average distance between intersection points. There is a remarkable variation when the number of available nodes changes, as the running times reduce significantly when five VMs are used instead of only one. For instance, when the intersection points are spaced by 5 km and one virtual machine runs the GA, the average processing time is 184 min and standard deviation of 70 min to estimate alignments to HPTs, and 180 min with standard deviation of 101 min adequate to HST. With five VMs

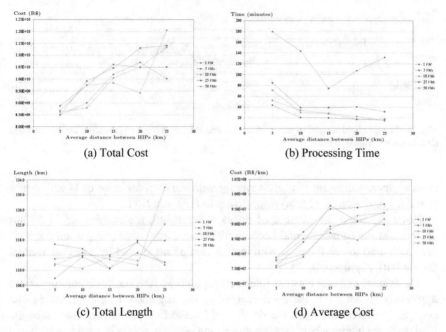

(a) Total Cost

(b) Processing Time

(c) Total Length

(d) Average Cost

Fig. 9 Average values of the estimated alignments for HST between Rio de Janeiro and Juiz de Fora

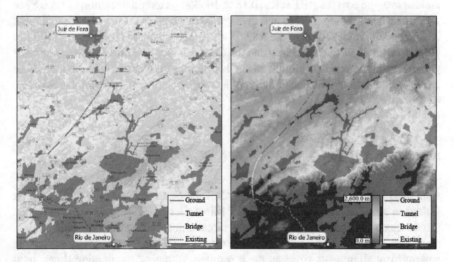

Fig. 10 Estimated lowest cost alignment over land use (left) and DEM (right) to operate HST between Rio de Janeiro and Juiz de Fora

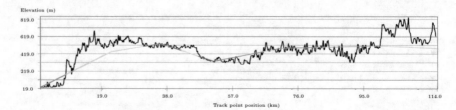

Fig. 11 Longitudinal section of the estimated lowest cost alignment to operate HST between Rio de Janeiro and Juiz de Fora

running, these values are 36 min on average with standard deviation of 24 and 85 min (standard deviation of 42 min) respectively to HPT and HST.

However, when the number of nodes is increased, the running times do not decrease in the same rate and become higher in some cases (e.g., when 25 VMs are configured to run the code with intersection points spaced by 5 km on average for the HPT alignments).

The average distance between intersection points and the number of running nodes apparently do not affect the results regarding the alignment length. Therefore, the cost per km increases when the number of intersection points increases since the total costs are higher when the average spacing between points is greater, which can be explained by refined estimations of earthwork volumes. In the worst case, the highest average cost for HPT is R\$10.18 $\times$ 10^7/km (standard deviation of R\$ 0.87 $\times$ 10^7/km) when running 5 VMs and average distance between intersection points of 25 km, and the values for HST are R\$9.67 $\times$ 10^7/km (standard deviation of R\$0.91 $\times$ 10^7/km) with 25 virtual machines and the same number of points. The average costs per kilometer to build new alignments in this case are similar due to the mountainous terrain, which require the construction of a set of tunnels near the coastal area of Rio de Janeiro.

4.3   Case 2: Campinas-Poços de Caldas

Campinas is 90 km away from the capital of the State of São Paulo and Poços de Caldas is a touristic town in the State of Minas Gerais. They are separated by flat terrain close to Campinas that becomes mountainous near Poços de Caldas, with small surrounding cities in their path. The results of the parallel GA when estimating new railway alignments between them resulted in total costs, running times, total length and costs per kilometer depicted in Figs. 12 and 15 regarding the estimated alignments suitable to operate HPT and HST, respectively. Figures 13, 14, 16 and 17 illustrate the horizontal and vertical profile of the lowest cost alignments regarding the respective technologies.

As can be seen, the variation of total costs is small across the average distance between intersection points and number of nodes for both technologies. The average

Parallel Genetic Algorithm and High Performance Computing ... 175

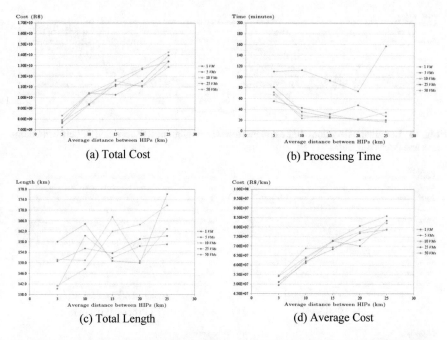

(a) Total Cost

(b) Processing Time

(c) Total Length

(d) Average Cost

Fig. 12 Average values of the estimated alignments for HPT between Campinas and Poços de Caldas

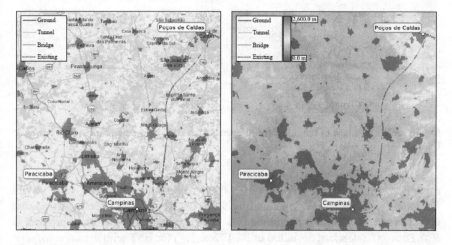

Fig. 13 Estimated lowest cost alignment over land use (left) and DEM (right) to operate HPT between Campinas and Poços de Caldas

Fig. 14 Longitudinal section of the estimated lowest cost alignment to operate HPT between Campinas and Poços de Caldas

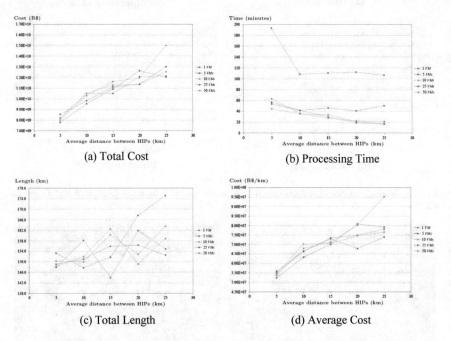

(a) Total Cost   (b) Processing Time

(c) Total Length   (d) Average Cost

Fig. 15 Average values of the estimated alignments for HST between Campinas and Poços de Caldas

minimum cost for HPT is R\$7.22 $\times$ 10^9 (standard deviation of R\$0.67 $\times$ 10^9) when the intersection points are separated by 5 km on average (24 intersection points) and 10 VMs are available to run the GA. On the other hand, the minimum cost is R\$7.78 $\times$ 10^9 (standard deviation of R\$0.75 $\times$ 10^9) for HST with the same average distance between HIP and 10 available VMs. The estimations tend to be similar to the previous case as the total costs increase when the average distance between intersection points raise, but do not vary significantly across the number of available nodes.

There is a remarkable reduction in running times when more than one VM is used to execute the Genetic Algorithm, which do not change significantly when more than five nodes are turned on. For instance, the average elapsed time to solve the RAO

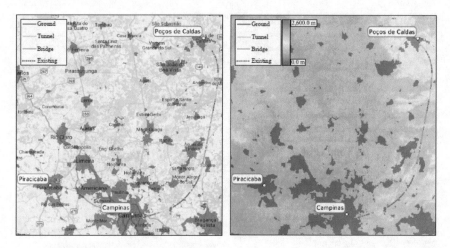

Fig. 16 Estimated lowest cost alignment over land use (left) and DEM (right) to operate HST between Campinas and Poços de Caldas

Fig. 17 Longitudinal section of the estimated lowest cost alignment to operate HST between Campinas and Poços de Caldas

problem for HPT when the intersection points are spaced by 25 km on average is 157 min (standard deviation of 50 min) with one VM running and, on the other hand, the convergence is reached in 28 min on average (standard deviation of 14 min) if 5 nodes are running in the high performance computing environment. For HST, the largest difference is observed when the intersection points are separated by 5 km, with elapsed time of 193 min on average (standard deviation of 83 min) for one single running node, and 57 min (standard deviation of 35 min) and 63 min (standard deviation of 46 min) minutes with 5 and 50 VMs respectively.

The length of the estimated alignments gradually increases when the number of intersection points reduces, i.e., when the average distance between them increases. Despite the variation of the estimated distances when running the GA with a different number of nodes, these values do not affect the average cost per kilometer as they are relatively small compared to the total estimated costs.

The obtained costs per kilometer are lower compared to the previous case as the minimum average costs per km for HPT is R\$4.96 $\times$ 10^7/km (standard deviation of R\$0.71 $\times$ 10^7/km) and R\$5.24 $\times$ 10^7/km (standard deviation of R\$0.63 $\times$ 10^7/km)

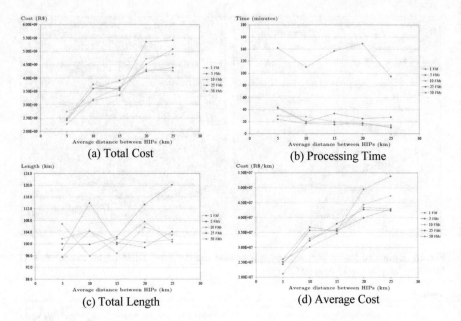

(a) Total Cost

(b) Processing Time

(c) Total Length

(d) Average Cost

Fig. 18 Average values of the estimated alignments for HPT between Araraquara and Ribeirão Preto

for HST, both for replications with intersection points spaced by 5 km on average and 5 VMs.

The estimated costs among the studied cases may differ due to the terrain configuration between Campinas and Poços de Caldas, with reduced earthwork costs. The length of tunnels and bridges is small in all the replications and, thus, their contribution to the total costs is smaller.

4.4   Case 3: Araraquara-Ribeirão Preto

The last case studied presents the solution to the RAO problem to estimate new alignments between Ribeirão Preto and Araraquara, two medium size cities in the State of São Paulo. Figures 18 and 21 show the results of total costs, the elapsed times to achieve these solutions, the total length and the cost per kilometer of the alignments suitable to operate HPT and HST respectively. Figures 19 and 20 illustrate the estimated lowest cost alignment to operate HPT over the land use and DEM database, and its longitudinal section, respectively. Similarly, Figs. 22 and 23 show the lowest cost alignment suitable to operate HSTs.

Despite the same trends compared to the other connections, the variance of total and average costs per km of the obtained alignments between Ribeirão Preto and Araraquara are closer to the results obtained between Campinas and Poços de Caldas.

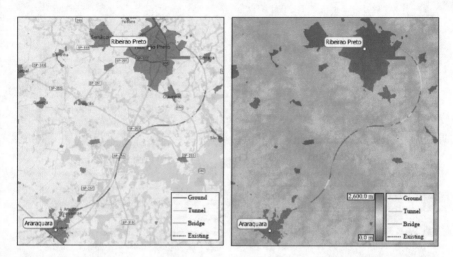

Fig. 19 Estimated lowest cost alignment over land use (left) and DEM (right) to operate HPT between Araraquara and Ribeirão Preto

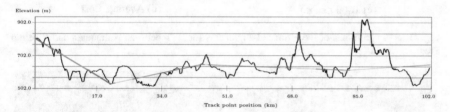

Fig. 20 Longitudinal section of the estimated lowest cost alignment to operate HPT between Araraquara and Ribeirão Preto

The minimum average total cost for HPT is R\$2.28 $\times$ 10^9 with standard deviation of R\$0.91 $\times$ 10^9 when the distance between intersection points is 5 km on average and 50 VMs are available. For HST these values are respectively R\$2.38 $\times$ 10^9 and R\$0.31 $\times$ 10^9 with an average distance between intersection points of 5 km and one running VM.

However, the elapsed times to retrieve the solutions are considerably smaller when one virtual machine is running compared to the performance of more than five nodes, while the results with a higher number of VMs is less deviated from the average than in the other studied cases. For HPT alignments, the minimum average processing time is 142 min with a standard deviation of 108 min when running one node and the intersection points are spaced by 5 km on average. When four nodes are added to solve the problem, the average running time reduces to 44 min (standard deviation of 18 min) given the same average distance between intersection points.

Besides, the minimum average computational time to obtain alignments for HST is 162 min (standard deviation of 57 min) when the intersection points are separated by 5 km, and one VM is used. Furthermore, the processing time to achieve the

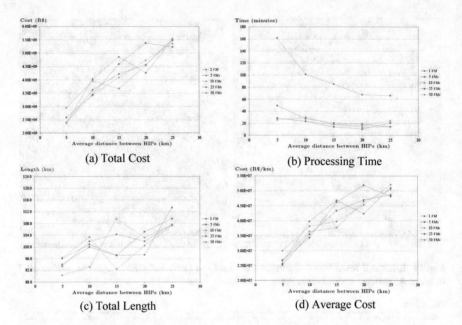

(a) Total Cost

(b) Processing Time

(c) Total Length

(d) Average Cost

Fig. 21 Average values of the estimated alignments for HST between Araraquara and Ribeirão Preto

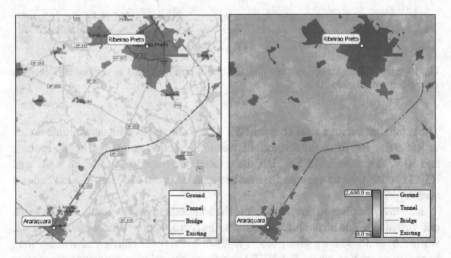

Fig. 22 Estimated lowest cost alignment over land use (left) and DEM (right) to operate HST between Araraquara and Ribeirão Preto

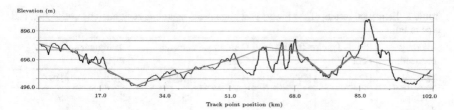

Fig. 23 Longitudinal section of the estimated lowest cost alignment to operate HST between Araraquara and Ribeirão Preto

solutions using 5 nodes is reduced to 50 min on average with a standard deviation of 19 min.

The average length of the optimal solutions for HPT is 95.6 km (standard deviation of 11.6 km) when running one virtual machine with average distance between intersection points of 5 km. The obtained alignments for HST lead to an average length of 93.4 km with a standard deviation of 8.2 km under the same conditions.

The average cost per kilometer to build HPT alignments is R\$2.67 $\times$ 10^7/km with standard deviation of R\$0.67 $\times$ 10^7/km obtained by running the GA in 50 nodes with average distance of 5 km between HIPs, which is significantly smaller than the results obtained in the previous cases. On the other hand, the obtained value for HST equals R\$2.52 $\times$ 10^7/km on average with standard deviation of R\$0.87 $\times$ 10^7/km given an average distance between HIPs of 5 km and 5 VMs.

4.5 Comparative Analysis

This section aims to compare the results of the parallel Genetic Algorithm applied to the intercity connections previously described. Table 2 summarizes the results of the estimated alignments of minimum total cost in each studied cases.

The results show that the lowest cost solutions to the RAO problem are when the average distance between intersection points is 5 km as it provides the most refined estimations of earthwork volumes and, thus, lower values to the most representative cost item among the overall estimated alignment cost. The number of Genetic Algorithm generations to achieve the solutions is not influenced by the railway technology and the running times are proportional to the number of iterations.

The minimum total costs are close to each other when the technologies are compared within the same case studied. However, the costs tend to reduce for both technologies when the connections of Rio de Janeiro-Juiz de Fora and Campinas-Poços de Caldas are compared to Araraquara-Ribeirão Preto since the terrain between the cities becomes flatter in Case 3.

The total estimated cost is not directly associated with the distance since the total length of the alignments for both technologies between Rio de Janeiro and Juiz de Fora is smaller than the Araraquara-Ribeirão Preto case. While the respective length

Table 2 Estimated results of the lowest cost alignments regarding each intercity connection and type of train

Intercity connection	Rio de Janeiro-Juiz de Fora		Campinas-Poços de Caldas		Araraquara-Ribeirão Preto	
Technology	HPT	HST	HPT	HST	HPT	HST
Average distance between HIPs (km)	5	5	5	5	5	5
Number of HIPs	19	19	24	24	14	14
Number of generations	340	877	645	1081	755	648
Running time (min)	23	58	67	93	47	35
Total cost (R\$ $\times 10^9$)	7.00	7.67	6.19	6.23	1.51	1.87
Length (km)	111.1	106.7	146.3	149.1	94.5	87.9
Cost per kilometer	62.98	71.87	42.33	41.75	15.94	21.22
Infrastructure cost (R\$ $\times 10^9$)	0.17 (2.4%)	0.45 (5.9%)	0.23 (3.7%)	0.63 (10.1%)	0.15 (9.9%)	0.37 (19.8%)
Earthwork cost (R\$ $\times 10^9$)	0.63 (9.0%)	0.57 (7.4%)	0.71 (11.5%)	0.68 (10.9%)	0.25 (16.6%)	0.37 (19.8%)
Tunnel cost (R\$ $\times 10^9$)	4.12 (58.9%)	4.48 (58.4%)	2.64 (42.6%)	3.22 (51.7%)	0.51 (33.8%)	0.47 (25.1%)
Bridge cost (R\$ $\times 10^9$)	1.42 (20.3%)	1.44 (18.8%)	2.02 (32.6%)	1.1 (17.7%)	0.45 (29.8%)	0.47 (25.1%)
Expropriation cost (R\$ $\times 10^9$)	0.02 (0.3%)	0.02 (0.3%)	0.03 (0.5%)	0.03 (0.5%)	0.02 (1.3%)	0.01 (0.5%)

for HPT and HST in the first case is 111.1 km and 106.7 km, and the total costs are R\$7.00 $\times 10^9$ and R\$7.67 $\times 10^9$ respectively, the third case resulted in 94.5 km and 87.9 km for the respective technologies with total cost of R\$1.51 $\times 10^9$ and R\$1.87 $\times 10^9$. The terrain configuration considerably impacts the total cost to build the alignment regardless the railway technology since the tunnel costs represent around 64% of the total costs for both technologies in Case 1, and 37.2% and 27.6% in Case 3 for HPT and HST, respectively.

These results directly affect the average cost per kilometer of the alignments. While Case 1 resulted in R\$62.98 $\times 10^6$/km and R\$71.87 $\times 10^6$/km for alignments suitable to HPT and HST, respectively, the respective values of Case 2 are R\$42.33 $\times$

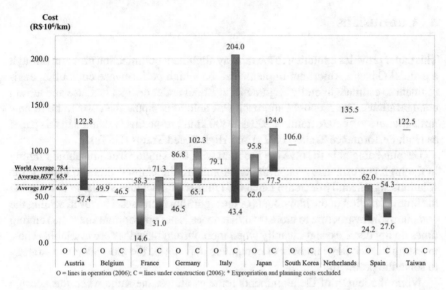

Fig. 24 Average costs per kilometer for construction of new HST rail alignments. *Source* Adapted from De Rus et al. (2009)

10^6/km and R\$41.75 × 10^6/km, and R\$15.94 × 10^6/km and R\$21.22 × 10^6/km in the third case. Thus, the average cost per km in Case 3 is 25.3% smaller than in the first one regarding the operation of HPT, and 29.5% for HSTs.

Given the estimated costs to build alignments appropriate to operate HPT and HST, the geometric parameters required to properly operate these technologies may not be the most relevant proxy since the earthwork, and tunnels and bridges costs are the most relevant cost items provided the terrain configuration.

In order to compare the estimated values, Fig. 24 presents the range of infrastructure construction costs required to operate High Speed Trains across different countries worldwide. Considering an average quotation of R\$3.10 per European currency unit (€) between 01/19/2015 and 05/19/2014 (BCB 2015), the minimum and maximum values observed refer to R\$14.6 × 10^6/km (France) and R\$204.3 × 10^6/km (Italy), respectively.

Despite the specific features of the Brazilian terrain and the premises of the proposed parallel Genetic Algorithm framework, Fig. 24 shows that the average cost values to build the railway alignments obtained in this paper are close to the international practice (De Rus et al. 2009). While the average construction cost per kilometer of HST alignments worldwide is R\$78.4 × 10^6/km, the average estimated values in Brazil are R\$65.9 × 10^6/km for HST and R\$63.6 × 10^6/km for HPT.

5 Conclusions

This paper provides a solution to the railway alignment optimization problem through a parallel Genetic Algorithm implemented on a high performance computing environment to estimate intercity alignments in Brazil under distinct land use and terrain configurations. The proposed approach was able to compute the costs to build new infrastructure to operate trains of 200 and 300 km/h, respectively defined in this paper as High Performance Trains (HPTs) and High Speed Trains (HSTs).

Despite being able to solve the problem with one single virtual machine running in the high performance computing environment, the elapsed times to achieve the solutions significantly decrease when the number of machines available to compute the fitness function of the individuals of the algorithm increases. However, scaling the computing infrastructure to more than five nodes is not appropriate since the running times do not decrease significantly when more virtual machines are available. Thus, investing in such large number of computers may not be the most adequate strategy to solve the problem.

While the length of the alignments remains almost the same when the average distance between intersection points is higher (because of slight differences in the number of horizontal and vertical curves), the total costs decreases with higher density of points (i.e., lower average distance between intersection points) since the tunnels and bridges are replaced by cutting and embankments.

Throughout the application of the model to three real-world cases, we showed that the alignment costs are not affected by the number of nodes available to run the Genetic Algorithm and that the high performance computing infrastructure affects only the running times and not the problem solution itself.

As expected, the alignments adequate to operate HPT and HST have less number of curves in flat terrains such in the studied cases of Campinas-Poços de Caldas and Araraquara-Ribeirão. However, in these scenarios the shape of the alignments is different between train technologies. On the other hand, the estimated alignments between Rio de Janeiro-Juiz de Fora are similar among train technologies because the mountainous terrain between them require more tunnels and bridges regardless the type of train.

Despite the capabilities of the parallel GA framework to solve the RAO problem, it can still benefit from technical improvements as the total estimated costs may include the values of operating costs on a two-level approach of RAO solutions followed by a train performance simulation. Moreover, methodological research would be carried to investigate the performance of a concurrent computing approach compared to the parallel environment described in this paper.

Acknowledgements The authors thank the reviewers for providing helpful comments and suggestions that improved the presentation of the paper. They acknowledge the use of computational facilities of the Laboratory of Advanced Scientific Computation of the University of São Paulo (LCCA-USP) between 2014 and 2015. The first author also acknowledges Brazil's CNPq grant 142417/2010-6 for the financial aid through a doctoral scholarship.

References

AREMA (2003) Railway track design. In: Practical guide to railway engineering, chapter 6. American Railway Engineering and Maintenance-of-Way Association, Maryland, USA

BCB (2015) Banco Central do Brasil - Taxas de Câmbio. http://www4.bcb.gov.br/pec/taxas/port/ptaxnpesq.asp?id=txcotacao. Accessed on 19/01/2015

Chen JX, Guo YY, Lv MX (2011) An adaptive genetic algorithm based on multi-population parallel evolutionary for highway alignment optimization model. Appl Mech Mater 58:1499–1508

Cheng J-F, Lee Y (2006) Model for three-dimensional highway alignment. J Transp Eng 132:913–920

Chew EP, Goh CJ, Fwa TF (1989) Simultaneous optimization of horizontal and vertical alignments for highways. Transp Res Part B Methodological 23:315–329

Crainic TG, Davidović T, Ramljak D (2012) Designing parallel meta-heuristic methods. In: Despotovic-Zrakic M, Milutinovic V, Belic A (eds) Handbook of research on high performance and cloud computing in scientific research and education, chapter 11. IGI Global, Québec, Canada

De Rus G, Barrón I, Campos J, Gagnepain P, Nash C, Ulied A, Vickerman R (2009) Economic analysis of high speed rail in Europe. Fundación BBVA, Spain

De Smith MJ (2006) Determination of gradient and curvature constrained optimal paths. Comput-Aided Civ Infrastruct Eng 21:24–38

Dolinayova A, Danis J, Cerna L (2018) Regional railways transport—effectiveness of the regional railway line. In: Fraszezyk A, Marinov M (eds) Sustainable rail transport. Lecture notes in mobility

Easa SM (1988) Selection of roadway grades that minimize earthwork cost using linear programming. Transp Res Part A Gen 22:121–136

Easa SM, Mehmood A (2008) Optimizing design of highway horizontal alignments: new substantive safety approach. Comput-Aided Civ Infrastruct Eng 23:560–573

Gupta M, Deep K (2009) A state-of-the-art review of population-based parallel meta-heuristics. In: World congress on nature & biologically inspired computing. TEEE, Coimbatore, India, pp 1604–1607

Hodas S (2014) Design of railway track for speed and high-speed railways. Procedia Eng 91:256–261

Howard BE, Bramnick Z, Shaw JF (1968) Optimum curvature principle in highway routing. J Highw Div 94:61–82

IBGE (2014) Instituto brasileiro de geografia e estatística - mapa mural de cobertura e uso da terra. http://www.metadados.geo.ibge.gov.br/geonetwork_ibge/srv/por/metadata.show?id=11079&currTab=simple. Accessed on 25/01/2014

Jha MK (2003) Criteria-based decision support system for selecting highway alignments. J Transp Eng 129:33–41

Jha MK, Schonfeld P (2000) Integrating genetic algorithms and geographic information system to optimize highway alignments. Transp Res Rec J Transp Res Board 1719:233–240

Jha MK, McCall C, Schonfeld P (2001) Using gis, genetic algorithms, and visualization in highway development. Comput-Aided Civ Infrastruct Eng 16:399–414

Jha MK, Schonfeld P, Jong JC, Kim E (2006) intelligent road design, vol 19. WIT Press, Boston, MA, USA

Jha MK, Schonfeld P, Samanta S (2007) Optimizing rail transit routes with genetic algorithms and geographic information system. J Urban Planning Dev 133:161–171

JPPF (2014) Java parallel programming framework—jppf. http://www.jppf.org/. Accessed on 18/04/2014

Kang M, Jha MK, Schonfeld P (2012) Applicability of highway alignment optimization models. Transp Res Part C Emerg Technol 21:257–286

Kazemi SF, Shafahi Y (2013) An integrated model of parallel processing and PSO algorithm for solving optimum highway alignment problem. In: Proceedings of the 27th European conference on modelling and simulation. European Council for Modeling and Simulation, Aalesund, Norway, pp 551–557

Kim E, Jha MK, Son B (2005) Improving the computational efficiency of highway alignment optimization models through a stepwise genetic algorithms approach. Transp Res Part B Methodological 39:339–360

Lai X, Schonfeld P (2016) Concurrent optimization of rail transit alignments and station locations.
Urban Rail Transit 2:1–15

Li W, Pu H, Zhao H, Liu W (2013) Approach for optimizing 3d highway alignments based on
two-stage dynamic programming. JSW 8:2967–2973

Li W, Pu H, Schonfeld P, Zhang H, Zheng X (2016) Methodology for optimizing constrained
3-dimensional railway alignments in mountainous terrain. Transp Res Part C Emerg Technol
68:549–565

Li W, Pu H, Schonfeld P, Yang J, Zhang H, Wang L, Xiong J (2017) Mountain railway alignment
optimization with bidirectional distance transform and genetic algorithm. Comput-Aided Civ
Infrastruct Eng 32:691709

Maji A (2017) Optimization of horizontal highway alignment using a path planner method. WIT
Trans Built Environ 176:81–92

MySQL (2014) Open source database—MySQL. http://www.mysql.com/. Accessed on 10/07/2014

OECD (1973) Optimisation of road alignment by the use of computers. Organisation for Economic
Co-operation and Development, Paris, France

Pu H, Zhang H, Li W, Xiong J, Hu J, Wang J (2018) Concurrent optimization of mountain railway alignment and station locations using a distance transform algorithm. Comput Ind Eng
127:1297–1314

Samanta S, Jha MK (2011) Modeling a rail transit alignment considering different objectives. Transp
Res Part A Policy Pract 45:31–45

Schonfeld P, Kang M, Jha MK, Karri GAK (2007) Improved alignment optimization and evaluation.
Technical report, Maryland, USA

TAV (2014) Trem de Alta Velocidade no Brasil. http://www.antt.gov.br/index.php/content/view/
5448/Trem_de_Alta._Velocidade_TAV.html. Accessed on 20/09/2014

USGS (2014) Earth explorer. http://earthexplorer.usgs.gov/. Accessed on 25/08/2014

The Use of Public Railway Transportation Network for Urban Intermodal Logistics in Congested City Centres

Lino G. Marujo, Edgar E. Blanco, Daniel Oliveira Mota and João Marcelo Leal Gomes Leite

Abstract The Mega-cities around the world are experiencing a rapid growth demanding more and more services and products in urban areas, which are often very dense and congested. The traditional road-based logistics strategies have been inadequate in dealing with large restricted delivery operations. Challenges arise due to legislation that restricts the travel of trucks inside the city centres, or to the increasing number of medium size vehicles in the streets. This paper shows a new method by comparing two options of delivering products in the city centres, one using traditional road-based delivery, and another with a hub-spoke model using the public railway transit system. Data analysis of the city's Master Transportation Plan shows a low level of utilization of the public railway transit system, and through this finding, we analysed the inter-modal freight transportation in such urban areas. To cope with transit time and cost, we developed a model, assessing the sensitivity and opportunities by carrying out analysis in a real case study. There is evidence that the model is sensible for the proposed congestion factor in the transit time as it can be a way to improve the service level of the deliveries inside the city centres, decreasing the number of medium sized trucks needed. The results show that it is possible to use inter-modal transportation when the road-based distribution operations suffer from a certain level of congestion in the haulage and last-mile stage.

L. G. Marujo (✉) · J. M. L. G. Leite
Industrial Engineering Program, Universidade Federal Do Rio de Janeiro, Rio de Janeiro, RJ 21941-909, Brazil
e-mail: lgmarujo@ufrj.br

J. M. L. G. Leite
e-mail: joaomarceloleite@gmail.com

L. G. Marujo · E. E. Blanco
Megacity Logistics Lab, Center for Transportation & Logistics, Massachusetts Institute of Technology, Cambridge, MA, USA
e-mail: eblanco@mit.edu
URL: https://www.megacitylab.mit.edu

D. O. Mota
Industrial Engineering Department, POLI/USP, Sao Paulo, SP, Brazil

© Springer Nature Switzerland AG 2020
M. Marinov and J. Piip (eds.), *Sustainable Rail Transport*, Lecture Notes in Mobility,
https://doi.org/10.1007/978-3-030-19519-9_6

Keywords Public railway network · Inter-modal operations · Urban logistics · Congestion factors · Megacity

1 Introduction

The Mega-cities[1] around the world are facing a rapid growth in their markets. An increasing number of new customers are demanding more and more services and products in urban areas, mainly in emerging markets, which are often very dense and congested. The process of fulfilling these needs is denominated "City Logistics". One critical component is the urban distribution of products, predominantly road-based, which contributes to traffic congestion and is negatively impacting the sustainability dimensions (environmental and financial).

In urban areas, the logistics operations show specific characteristics differentiating them from general logistics activities (Barceló et al. 2007). For example, the parking process is restricted to a few available areas and time to perform deliveries is shorter than at the city outskirts. The relevance of urban freight transportation can also be shown by the last-mile distribution cost within the freight transportation chain. About 40% of the costs are related to pick-up and delivery operations for the total door-to-door cost (Taniguchi and Thompson 2004).

Nevertheless, the intermodal freight chain in urban contexts is little explored compared to interurban distribution operations. Intermodal networks can be characterized by nodes, representing transfer or transhipment points and links, and the possible routes and flows of goods on a network (Crainic and Kim 2007). Hence, the inter-relations between different possible modes of transportation to deliver products to customers, is not well analysed in the urban context.

The general objective of this paper, therefore, is to provide an innovative framework for analysing the intermodal operations inside the high-density city centres using alternative schemes of distribution.

The specific aims of this paper are twofold. Firstly, we set out to analyse the utilization of the public railway system in general and identify the opportunities within its capacity. Secondly, we aimed to establish an intermodal cost and time model urban, intermodal operations that are subject to congestion patterns in the stem haul and last-mile section of the delivery process. Both objectives are based on the data available from the public railway systems and demographics patterns of a Mega-city.

The remainder of the paper is organized as follows. Section one reviews the current literature on intermodal logistic operations. Section two discusses the research motivation, the methodology and presents the developed models. Section three presents the numerical example for the public railway transportation system analysis and the intermodal operation scheme assessment for the entire city. Section four presents the discussions and concludes the paper.

---

[1]Cities with more than 10 million inhabitants.

Recently, Behrends (2012) stated a relationship between the urban transport system and intermodal transport strategy using the interface of road-rail modes. The aim was to identify possible strategic actions, taken by the local stakeholders, to improve the competitiveness and environmental benefits of rail freight. While pointing out some policies and actions, no explicit model was provided.

Some cities have adopted policies relating to access restriction schemes for certain types of cargo vehicles when entering city centres. This is an attempt to promote the efficiency of mobility and mitigate traffic jams. In this environment, several challenges are faced by deliverers including lack of space, the protection of activities of the citizens and their social-health dimensions. Cargo vehicles are restricted commonly by the time of day of access, by allowed license plates per weekday, by vehicle size and type of operations (Alessandrini et al. 2012).

Diziain et al. (2013), and Taniguchi and Nemoto (2008) have reported some initiatives about the use of intermodal freight transport in urban areas. These relate to the combination of road and rail freight transportation and road and waterways transportation. These case studies focus on garbage collection and transportation of rubbish to a disposal area in Japan, and crops and chemistry in France. Many challenges were highlighted such as the development of specific equipment, for example, containers, for these operations which required subsidies from the government for the initial investment. The authors have stated that intermodal transport of goods can be an option even on short distances. However, the model needs existing logistics facilities, located in the urban centre, adequate transport mode infrastructure (e.g. railways, waterways, etc.), high level of roads and public policy support. Thinking about how a train can carry people and goods, Kelly and Marinov (2017) suggested various designs for the interior of the metro. The authors concluded that is possible to have interior designs which allows good and people transport.

Marinov et al. (2013) have explored the idea of using the existing urban light rail networks for distributing goods in cities. They argued that an efficient distribution of goods by rail will have significant economic and environmental benefits and also minimize traffic congestion and greenhouse gas emissions, along with traffic-generated noise pollution. But the counter side is that it requires huge amount of investments. They analysed six case studies around Europe and stated whether or not an urban rail freight operation is feasible. Following the same idea, Singhania and Marinov (2017) built a simulation model for analysing the utilization levels of a section of a railway line. The authors evaluate the impacts of inclusion of new freight trains. The results proved the viability to include extra freight trains without prejudice to current passenger train timetables. Dampier and Marinov (2015) cited in their paper one of most important benefits of using metropolitan railways for freight transport is accident reduction, with a lower casualty rate.

Moreover, Janic (2007) developed a model for calculating the combined intermodal and road internal and external costs involved, taking into account the costs of the social and environmental impacts. Therefore, the model was implemented with a regional freight perspective; not in an urban context, not being affected by population density, nor congestion (another example is Hanssen et al. 2012). Motraghi and Marinov (2012) created an event based simulation model using ARENA to evaluate the use of urban rail for freight transport. The model allows users to analyse the actual situation and provide alternatives to improve rail system utilization. They used, as example, the Tyne and Wear Metro system in the North East of England.

A study of intermodal freight networks have shown a considerable contribution to the transportation carbon footprint analysis and possible mitigation strategies (see Craig et al. 2013). Craig et al. (2013) derives an expression from the work of Niérat (1997), with the objective of determining the maximum intermodal operation service area, around an origin or a destination, to establish when to use an intermodal operation or a road-based one.

A more generalized intermodal operation is analysed in the work of Smilowitz and Daganzo (2007), where they formulate a general model for a parcel collection, intermodal long-haul, delivery operations, based on the continuous approximation approach (Daganzo 2005). The model was also implemented on a regional perspective.

Another factor, the presence of congestion and its impacts in the logistics activities, has been studied by authors in the classical area of vehicle routing models, but none of these are tackling intermodal operations. They are considering the "green" factors of this activity, analysing the costs and the restrictions as well as the environmental impacts (see Crainic et al. 2004; van Woensel and Cruz 2009; Browne and Gomez 2011).

Figliozzi (2010) has developed an approach based on numerical experiments with real-world data to understand the impacts of congestion in the reliability of logistics operations. He has also analysed the increasing costs, time and distance borne by the carriers. With this cost and structure analysis, he could categorize the distribution system into three different groups, with many stops, fewer stops but inside a large delivery area, and located farther from the depot. The travel time was the fundamental factor to worsen the impact on congestion.

The structure of the last-mile delivery problem is analysed by Novaes et al. (2000). They presented a model of setting district boundaries for fleet planning, taken into account the capacity of the trucks, the time restrictions and the shape of each region of service delivery. These concepts are also used in this paper.

2 Conceptual Framework

2.1 Problem Statement

According to the data analysed, there exists a lack of utilization in public railway transportation networks. The railway system is usually overloaded during the peak-hours. However, in the middle of the day, between peak-periods, the system is sub-utilized. This capacity could be reverted to transport freight, either in a mixed way with passengers or with dedicated transportation units in a segregated schedule.

The traditional delivery method uses trucks from the distribution center to the client location, in a door-to-door schema. It is called a road-based system. Trucks travel the entire line haul inside the delivery zone, attending the customers. The suggested new method combines truck and railway systems to deliver goods to different parts of the city. It uses an intermodal scheme to perform the operations. It begins in a distribution center, where the first-mile (collection of the goods) is performed by trucks. After a transhipment operation, the product is transported by the public railway system to the next destination, the rail line-haul. Finally, transhipment is made to the last-mile delivery vehicles on smaller vehicles, where the products are delivered to the clients. In this work, we have studied a dataset comprising of a Mega-city population density and transit network to estimate the capacity of a railway transit system for freight. We have developed a model to study cost and time parameters for capacity analysis in an intermodal operation.

2.2 Methodology

Our first step was to analyse the city's Master Transportation Plan data to identify the opportunities and existing gaps in the use of the public railway transportation networks. We focussed on the peak-hour identification and the percentage of usage of the public railway system in general and the light rail system in particular. The rail system as an option, but also any transit system with dedicated lanes (for a complete classification of that, see Vuchic 2007). It is important to identify the most relevant period of usage and the critical rail branch for commuting to the city centre, the occupation of lines and stations, as well as the available capacity that could be used for freight.

The second step was to compare the road-based delivery system with the intermodal system. To make this comparison, it was necessary to evaluate the customer service level and the operational cost of each option. The customer service level was measured as the transit time between the distribution center and the customer destination. The cost and transit time evaluation of both operations, shown in Fig. 1, should comprise overall operation steps, including collecting, handling, transporting and delivering the goods to the clients. Both operations have the same initial point, that is the satellite distribution center on the outskirts of the city (Crainic et al. 2004). An important input for transit time calculation is the traffic jam level between the distribution center and the delivery zone and inside the delivery zone. Both models calculate the transit time and total cost of traffic jam levels as one inputs.

After comparing road-based and intermodal service levels and costs using current data, the model allows for elaborate as-if analysis. It is possible to evaluate the impact of traffic jam level changes and evaluate performance level of each delivery system from this model.

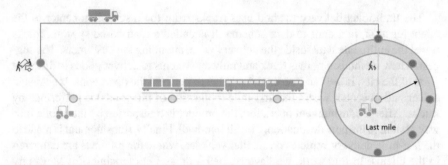

Fig. 1 Intermodal and road-based operations schema, showing the origin of the products and the destination delivery service area, called here as last-mile area. The delivery operation to the service area can be made by truck directly or using an intermodal operation

2.3 Road-Based Logistic System

As it was mentioned in the previous section, to evaluate the road-based logistics system required evaluation of customer service levels (measured as total transit time) and total cost. Both models (transit time and cost) are divided in two steps:

1. Line-haul transportation: includes all operations from the distribution center loading until arriving at the beginning of the delivery zone;
2. Last-mile transportation: a set of deliveries to the customers located inside the determined zone, with a volume v_i to attend each customer.

To calculate the transit time and the cost, we used approximate formulas to estimate the vehicle travelled distances and times along the route according Daganzo (2005), Novaes et al. (2000).

2.3.1 Transit Time Model

The Road-based logistic system transit time (T_{RB}) has two components:

1. line-haul transportation transit time until the last-mile zone (T_L);
2. last-mile transit time (T_z),

$$T_{RB} = T_L + T_z, \tag{1}$$

The line-haul transit time (T_L) can be expressed as:

$$T_L = \frac{2.d_{ik}.k_L}{v_L}(1 + \gamma_L) \tag{2}$$

where:

- k_L denotes the dimensionless factor for line-haul distance in Euclidean metric,
- d_{ik} is the distance from origin i to point of entrance into the zone delivery k ($D_L = 2 \cdot d_{ik} \cdot k_L$ denotes the line-haul distance),
- v_L is the free-flow velocity of line-haul transportation,
- γ_L is the traffic jam factor (Sheffi 1985) for the line-haul transportation.

According to Sheffi (1985) the transit time between two points i and k is affected by the traffic jam and can be calculated as:

$$t_{ik} = \frac{d_{ik}}{v_L} = t_{ik}^0 \left[1 + \alpha \left(\frac{f_{ik}}{c_{ik}} \right)^\beta \right] = t_{ik}^0 [1 + \gamma_{(.)}], \qquad (3)$$

where:

- t_{ik} is the transit time between i and k,
- t_{ik}^0 tik is the free flow transit time of the link i, k,
- f_{ik} denotes the flow on the link,
- c_{ik} is the capacity of the link i, k,
- α and β are traffic jams parameters,
- γ is the traffic jam factor ($\gamma = \alpha(f_{ik}/c_{ik})^\beta$) with $\gamma \in [0, 1]$.

The last-mile transit time can be determined by:

$$T_z = \left[\frac{2.r_{kj} + \frac{k_z}{\sqrt{\delta}}}{v_z} \frac{W_z}{\beta_i} \right] (1 + \gamma_z) = \left[\frac{k_z}{v_z . \sqrt{\delta}} + \delta . A_k . t_s \right] (1 + \gamma_z). \qquad (4)$$

where:

- r_{kj} is the radios of the delivery zone,
- k_z is the dimensionless factor for the delivery zone distance,
- δ represents the density of customers to be served, when the locations of clients follow a homogeneous 2-dimensional point process (e.g. Poisson). In this paper we assume that the density δ of visiting points over the region is constant. According to Daganzo (2005) and Novaes and Graciolli (1999), the distance of deliveries inside a delivery zone can be approximate by $k_z \cdot \delta^{-0.5}$,
- v_z is the free-flow velocity for the last-mile transportation,
- W_z is the capacity of the vehicle,
- β_i is the average demand per point of delivery,
- A_k represents the area of the zone of delivery k,
- t_s the stop time per deliver,
- γ_z is the traffic jam factor (Sheffi 1985) for the last-mile transportation.

Let n be the number of visits performed in a tour. We assume that the displacement time between two clients within the district of area A (last-mile delivery) suffers under traffic congestion, so $\delta = n/A$. In our work, we were not interested in finding the best sequence of visiting points for each tour in the delivery service for a certain district, constrained by congestion or not (as one can see in Novaes et al. 2009 and others abundant literature as in Pillac et al. 2013).

The maximum number of stops attended per vehicle is $C = \frac{W_z}{\beta_i} = 2.\delta.w.L$, where the w and L are delivery zone dimensions of width and length, respectively.

2.3.2 Cost Model

As the transit time model, the Road-based logistic system cost (C_{RB}) has two components:

1. line-haul transportation transit time until the last-mile zone (C_L);
2. last-mile transit time (C_z),

$$C_{RB} = C_L + C_z, \tag{5}$$

The line-haul cost (C_L) and the last-mile cost (C_z) can be expressed as:

$$C_L = T_L.\frac{c_L^v}{W_L}, \tag{6}$$

$$C_Z = T_Z.\frac{c_Z^v}{W_Z}, \tag{7}$$

where:

- T_L *is the line-haul transit time,*
- c_L^v *represents the variable cost per time in the line-haul transportation,*
- W_L *is the capacity of the vehicle in the line-haul transportation,*
- T_Z *is the last-mile transit time,*
- c_z^v *represents the variable cost per time in the last-mile transportation,*
- W_z *is the capacity of the vehicle in the last-mile transportation.*

2.4 Multimodal Logistic System

To calculate both customer service level (measured as total transit time) and total cost, the operation is divided in five steps:

1. First-mile transportation: collecting operation from origin i to the station l,
2. Train loading: handling operation at station l to load the volume to attend the customer zone i in the train,
3. Rail haulage transportation: transport the volume $\sum_i v_i$ from station k to the station l, the beginning of customers' zones,

4. Train unloading: handling operation at customer zone's to unload i volume at station k,
5. Last-mile transportation: a set of deliveries to the customers located inside the deter- mined zone, with a volume v_i to attend each customer.

We will use the same methodology applied in the last session (Daganzo 2005; Novaes et al. 2000).

2.4.1 Transit Time Model

The Intermodal logistic system transit time (T_{IM}) has five components:

1. first-mile transportation transit time (T_{FM}),
2. loading handling time (H_l),
3. rail-haul transportation transit time until the last-mile zone (T_R),
4. unloading handling time (H_u),
5. last-mile transit time (T_z).

$$T_{IM} = T_{FM} + H_l + T_L + H_u + T_z, \tag{8}$$

The first-mile collection operation transit time is calculated as:

$$T_{FM} = \frac{2.d_{il}.k_L}{v_{FM}}.(1 + \gamma_L), \tag{9}$$

where:

- d_{il} denotes the distance between the origin i to transhipment point l,
- k_{FM} denotes the dimensionless factor for first-mile distance in Euclidean metric,
- v_{FM} denotes the free-flow velocity of first-mile,
- γ_{FM} is the traffic jam factor.

The handling time in the loading and unloading transhipment points is:

$$H_l = h_l \cdot L_l, \tag{10}$$

$$H_u = h_u \cdot L_u, \tag{11}$$

where:

- h_l and h_u denote the capacity for loading and unloading handling the quantity of cargo per hour, respectively,
- L_l and L_u is the time consumed for the loading and unloading handling capacity, respectively.

The rail haulage transit time is defined as

$$T_R = \frac{d_{lk}}{v_R} + h_k.L_k, \tag{12}$$

where:

- d_{lk} is the distance from the loading point l until the unloading station k close to the delivery zone,
- $h_{(.)}$ and $L_{(.)}$ follow the same definition of the handling operations (Eqs. 10 and 11),
- v_R is the free-flow velocity of rail haulage transportation.

Based on the work of Pachl and White (2004) the average minimum line headway t_h of the rail-line in study was obtained by:

$$t_h = \sum_{i,j} \frac{t_{h,i,j} n_i n_j}{n^2} \tag{13}$$

where:

- $t_{h,l,j}$ is the minimum line headway adopted for train type j following train type i, for example, express service and regular,
- n_i, n_j are the number of trains for each type,
- n is the total number of trains.

With expression 13, one can determine the total capacity of trains in such branch and with the capacity of each railcar, derive the traction capacity per wagon in kilograms, $W_r = \psi \hat{w} l_r w_r$, once we have the average occupancy of the rail-line, ψ, in terms of passengers per square meters, pass/m^2, the weight profile of the population that uses this branch, $\hat{w}$, and the railcar dimensions l_r, w_r.

Finally, the last-mile transit time approximation function for the intermodal operation is determined in the same way as previously, by:

$$T_z = \left[\frac{2.r_{kj} + \frac{k_z}{\sqrt{\delta}}}{v_z} \times \frac{W_z}{\beta_i} \right] (1 + \gamma_z) \tag{14}$$

Thus, the total cost for intermodal operation (C_{IM}) is given by:

$$C_{IM} = T_{FM} \cdot c_{FM}/W_{FM} + H_l \cdot c_l/W_l + T_R \cdot c_R/W_R + H_u \cdot c_u/W_u + T_z \cdot c_z/W_z \tag{15}$$

where:

- $c_{(.)}$ is the cost per time for each operation: first-mile (FM), loading (l), rail-haul transportation (R), unloading (u) and last-mile (z),
- $W_{(.)}$ is the cost per time for each operation: first-mile (FM), loading (l), rail-haul transportation (R), unloading (u) and last-mile (z).

Table 1 Passengers boarding per branch, per day (adapted from SETRANS 2013b)

Branch	Boarding
Deodoro	256,725
Japeri	115,587
Santa Cruz	81,906
Saracuruna	54,766
Belford Roxo	27,663

3 Numerical Example

3.1 Analysis of the Public Railway System

Rio de Janeiro is a mega city in Brazil. The estimated population including neighbouring municipalities is about 12 million people. The region has more than 4 million vehicles. Rio de Janeiro has a very complex transportation system and suffer with traffic jams due to its geographical characteristics. Rio is a beach town with many mountains, and a poor public transport system.

The analysis of the public railway transportation system of the Rio de Janeiro city started with the tabulation of the Origin-Destination (OD) matrix. This OD matrix is based on 265.000 interviews made in 2012, resulting in about 22.600.000 daily trips.

The analysis of the OD matrix has shown only 24% of the population uses the public transportation system for their morning and evening commute to the central area of the city. Furthermore, only 31% of people use the railway to commute to the city Downtown, which indicates others modes of transportation for commuting (SETRANS 2013a, b).

The modal share evolution shows a 10% decrease in the usage of public transportation and an increase of 15% in the usage of private cars, adding to a 73% growth in the car fleet. Another important analysis is the percentage of total daily trips at each working hour (Fig. 2). For private cars, the peak hour is 7–8 a.m. with 8.00% and the less busy hour is 6–7 a.m. with 6.25% (21.9% less, comparing to peak hour). For public transit, the peak hour is 7–8 a.m., with 8.79% and the less busy is 13–14 with 6.12% (30% less, comparing to peak hour). This result shows a bigger lack of utilization for the public transport.

The profile of the most used branch in Rio de Janeiro City (Deodoro branch—Table 1) shows several opportunities for using the line as freight line, once the peak-periods of passengers are well defined, as shown in Fig. 3. We can see the travel demand to the city, which rises and falls over the course of a day. It characterizes a time-dependent travel demand pattern, and the operating day can be divided into pieces, as the rush periods and the off-peak periods (Fig. 4).

One point to be highlighted is the existing correlation between the low usage of public transportation and the growing usage of private cars in the cities. This, which

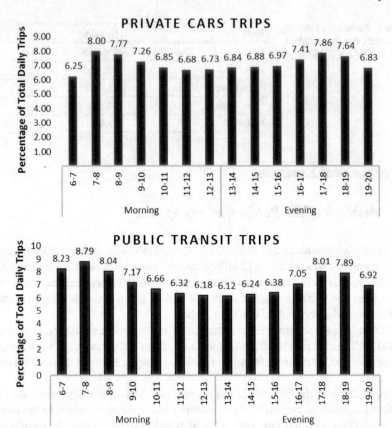

Fig. 2 Percentage of daily trips by hour of the day for private car users and for public transit users. The gap between the maximum and the minimum percentage is more relevant in the public transit trips than the one in the private car trips

leads to an increasing number of traffic-jams and under-utilization of the public transit system, especially out of peak periods.

Therefore, once the public railway system is already implemented and established, these systems could be used for carrying goods from the suburbs to the city in Mega-cities in a dedicated lane without interference or congestion.

In our analysis, the entire city was divided into 2250 small pieces, called generally as km^2, and for each km^2, the distance to each station was calculated. This produces a graphical structure of the service regions over the entire rail network, oriented by the nearest neighbour logic.

The demand for each km^2 was calculated as a function of the daily trash generation. This generation was related to the concentration of inhabitants in each region to find the estimated demand (COMLURB 2014).

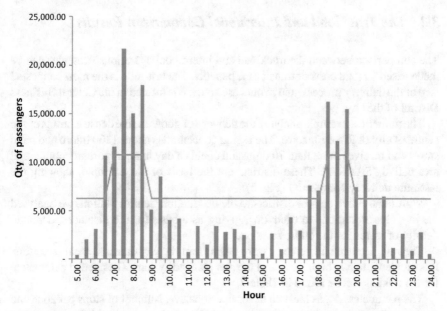

Fig. 3 Usage profile of the Deodoro rail branch, expressed as the quantity of passengers. The straight line represents the capacity of the rail line for a typical day. There are several underutilized periods between the peak ones

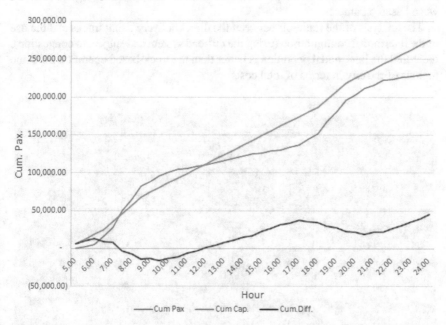

Fig. 4 Cumulative capacity and demand profile of the analysed rail branch per hour. The difference of the capacity and demand is plotted in the curve below

3.2   The Truckload and Intermodal Comparison Results

The comparison between the truckload and intermodal operations strategies can be made based on total costs or transit time benefits. The last-mile is the most congested step of the delivery process, and in our case study, it is located in the Central Business District of the city.

The numerical example analyses the delivery of goods in five dense districts of the centre of city of Rio de Janeiro. The region concentrates about 1200 micro and nano stores and receives more than 70 thousand clients a day in an important commercial area called "SAARA". These districts are the basis of our example, showing the last-mile delivery complexity (Fig. 5).

Within these five closest districts to the destination Central Station, we analysed the five closest stations to each km^2, giving us a basis for the distances, demand, density of stores for the entire rail system.

The rail distance and velocity were taken from the city data and local operators. It is divided into five branches, as in the real planned system. The needed parameters for the models can be found in Table 2.

The parameters A_z, as the area of service zone, N_z, Number of stops per zone and β_i, let be the average demand per stop and W_b, as the capacity of the railcar that was estimated based on the model of railcar and the load factor, come from the city analysis data described in the previous section, and the parameter c_h is a monthly wage based parameter.

The analysis of the trade-off between the direct delivery using trucks and the use of the intermodal configuration (using the railroad system and subject to congestion), identified the intermodal operation is better than the truck-based operation for some regions of the city in terms of total costs.

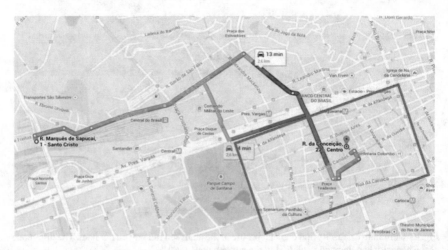

Fig. 5 Last-mile region characterization of the region called Saara in Rio de Janeiro's downtown (from: maps.google.com)

Table 2 Truckload and Intermodal model parameters and units (based on values provided by Janic (2007) and updated for the 2018 year)

Parameter	Description	Value	Unit
k_L	Dimensionless factor for linehaul distance	1.15	
c_L^v	Cost of linehaul per hour	45.10	$/hr
c_z^V	Cost of zone delivery per hour	25.72	$/hr
W_L	Capacity of linehaul vehicle	12,000.00	kg
W_z	Capacity of last mile vehicle	12,000.00	kg
k_z	Dimensionless factor for last mile	1.15	
t_z	Stop time per customer	0.60	hr
c_h	Handling cost per hour at origin station	4.39	$/hr
c_k	Handling cost per hour at end station	$1.2 \cdot c_h$	$/hr
h_l	Handling capacity in kg per hour	3000.00	kg/hr
c_r	Cost per railcar per km	0.77	$/km
ψ	Load factor of railcar per m^2	4.00	pax/m^2
$\bar{w}$	Average weight profile of the		
	Population (IBGE 2013)	67.00	kg/pax

Some critical points arose from the analysis. The handling time at the entrance and destination stations are a function of the coming capacity from the previous transportation mode in that chain. The simulation results have shown that the more controlled these activities are, the more the intermodal system gains advantage.

One first results was the capacity profile per day within the city. It is related to the minimal distance between each km^2 to the nearest station, and so, compared with the estimated demand. It shows the requirement to attend the regions, on a daily basis, is achieved by small vehicles, less than 12 tons (Fig. 6).

There is a specific point of trade-off in terms of cost per ton, as a result of the congestion's influence in the last-mile velocity, as stated by Fig. 7, for the critical branch analysis. This result is due to the dedicated lanes that the rail system has, avoiding the congested infrastructure in part of the delivery.

This can be understood in two fold. Firstly, the total cost per ton is less affected by this factor, which induces to a higher threshold of the velocity, because, in the intermodal operation, more cost to deal with handling the goods are needed. Therefore, for transit time-based analysis, the truck-based operation is more suitable once it can make fewer movements with the cargo.

The analysis of the total transit time, and how it varies with the velocity in the last- mile (most congested part of the delivery), shows that the intermodal schema is worthy in velocities bellow 17 km/h (Fig. 8).

The analysis of the entire system shows that there is a concentration of regions served by each station. Figure 9 shows the heat map for the regions (km^2) served by each station. Some stations serve a 427 km^2 population area.

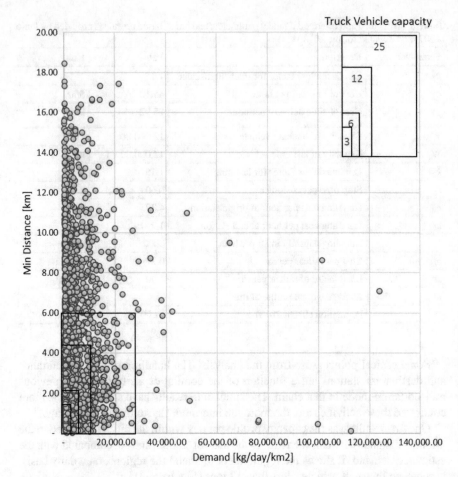

Fig. 6 Analysis of the capacity profile to serve each region in a day by one delivery vehicle

The variation of the congestion factor (γ) produces different sizes of service areas by the intermodal system to serve the city area. Figure 10 shows the service recommendation based in total cost for three different levels of congestion. While with $\gamma = 0\%$ only the west area of the city should use intermodal transport, with $\gamma = 50\%$ only the area close to the city should keep using road-based transportation. Figure 11 makes the same analysis based in transit time. While with $\gamma = 0\%$ only two limited areas should use intermodal transport (green area), with $\gamma = 0\%$ around 50% of the city area should use intermodal transport.

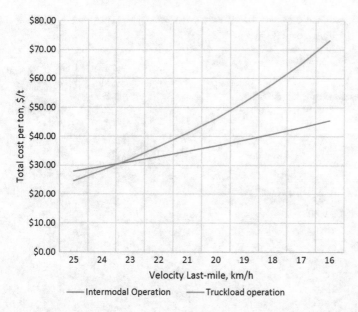

Fig. 7 Sensitivity analysis of the cost per ton delivered and the last-mile velocity affected by congestion. The velocity in the last-mile is affected by γ_L. The point of limit is around 23.5 km/h to intermodal operation becomes worthy

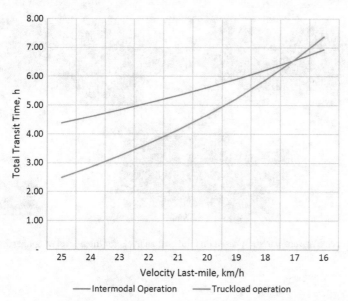

Fig. 8 Sensitivity analysis of the transit time and the last-mile velocity affected by congestion. The limit point to operate with truck-based delivery is about 17.5 km/h

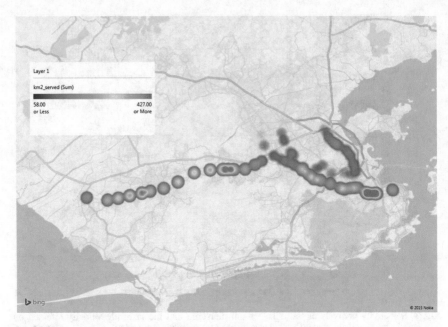

Fig. 9 Heat map for the regions served by each station of the rail system

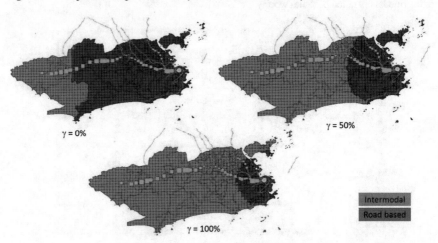

Fig. 10 Service regions variation based on total cost analysis for three level of congestion

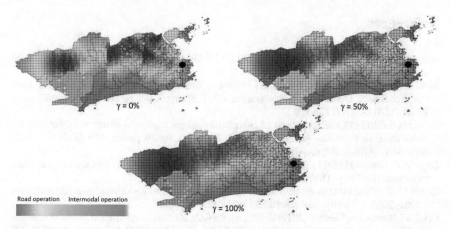

Fig. 11 Service regions variation based on transit time different analysis for three level of congestion

4 Conclusions

The model we developed was able to conduct the trade-off analysis between the use of a road-based system and an intermodal one making the deliveries inside the city centre. The possibility of gains in terms of reducing congestion in the cities' networks are discussed.

The simulation of each station and the surrounding 5 km^2 demonstrates the capacity of the existing rail system to manage the cargo that has a destination in the city centre. The demand served varies across the station as a function of the distance for each region to reach them.

The station profile to attend the demand is a function of the location and population density of the surroundings of each station. Some stations concentrate the demand to serve the km^2's, because they possess a privileged location. One region can be served by more than one station, but its attractiveness decreases in terms of cost, time and probability of congestion.

The system depends on a shared last-mile service system, in order to avoid under-utilization of the stations and the fleet, and the generation of undesired pipeline inventory. Future developments can be done to assess other commercial zones and the entire public railway system of a city, simulating the impact generated of the adoption of these strategies, and an analysis of the influence in the system of the different geographical patterns around the world, with their specific transit systems. Also, the inventory levels that can be created in each echelon of the distribution chain should be evaluated and constitutes a future pathway in this research.

Acknowledgements One of the authors (LGM) wishes to thank the financial support of the Brazilian National Scientific and Technological Council (CNPq), grant n° 248618/2013 − 0.
This study was financed in part by the Coordenação de Aperfeiçoamento de Pessoal de Nível Superior - Brasil(CAPES) - Finance Code 001.

References

Alessandrini A, Site PD, Filippi F, Salucci MV (2012) Using rail to make urban freight distribution more sustainable. Eur Transp 50:1–17

Barceló J, Grzybowska H, Pardo S (2007) Vehicle routing and scheduling models, simulation and city logistics. In: Dynamic fleet management, Ch. 8. Springer, US, pp 163–195. https://doi.org/10.1007/978-0-387-71722-7_8

Behrends S (2012) The urban context of intermodal road-rail transport—threat or opportunity for modal shift? In: Procedia—social and behavioral sciences, vol 39, pp 463–475. ISSN 18770428. https://doi.org/10.1016/j.sbspro.2012.03.122

Browne M, Gomez M (2011) The impact on urban distribution operations of upstream supply chain constraints. Int J Phys Distrib Logistics Manage 41(9):896–912

COMLURB (2014) Indicadores de coleta de resíduos sólidos urbanos - Município do Rio de Janeiro – 2002–2013, URL http://www.armazemdedados.rio.rj.gov.br

Craig AJ, Blanco EE, Sheffi Y (2013) Estimating the CO_2 intensity of intermodal freight transportation. Transp Res Part D: Transp Environ 22:49–53. ISSN 13619209. https://doi.org/10.1016/j.trd.2013.02.016

Crainic TG, Kim KH (2007) Intermodal transportation. In: Barnhart C, Laporte G (eds) Handbooks in operations research and management science, Chapter 8, vol 14. Elsevier, pp 467–537

Crainic TG, Ricciardi N, Storchi G (2004) Advanced freight transportation systems for congested urban areas. Transp Res Part C: Emerg Technol 12(2):119–137. ISSN 0968-090X. http://doi.org/10.1016/j.trc.2004.07.002

Daganzo CF (2005) Logistics systems analysis. Springer, Heidelberg

Dampier A, Marinov M (2015) A study of the feasibility and potential implementation of metro-based freight transportation in Newcastle upon Tyne. Urban Rail Transit 1(3):164–182. ISSN 2199-6679. https://doi.org/10.1007/s40864-015-0024-7. https://doi.org/10.1007/s40864-015-0024-7

Diziain D, Taniguchi E, Dablanc L (2013) Urban logistics by rail and waterways in France and Japan. In: 8th international conference on city logistics, France, pp 1–15

Figliozzi MA (2010) The impacts of congestion on commercial vehicle tour characteristics and costs. Transp Res Part E 46(4):496–506. ISSN 1366-5545. https://doi.org/10.1016/j.tre.2009.04.005

Hanssen TES, Mathisen TA, Jørgensen F (2012) Generalized transport costs in intermodal freight transport. In: Procedia—social and behavioral sciences, vol 54, pp 189–200. ISSN 18770428. https://doi.org/10.1016/j.sbspro.2012.09.738

IBGE (2013) Censo Demogr´afico Brasileiro de 2010. Technical report, IBGE. URL http://www.ibge.gov.br/home/estatistica/populacao/

Janic M (2007) Modelling the full costs of an intermodal and road freight transport network. Transp Res Part D: Transp Environ 12:33–44. ISSN 13619209. https://doi.org/10.1016/j.trd.2006.10.004

Kelly J, Marinov M (2017) Innovative interior designs for urban freight distribution using light rail systems. Urban Rail Transit 3(4):238–254. ISSN 2199-6679. https://doi.org/10.1007/s40864-017-0073-1. https://doi.org/10.1007/s40864-017-0073-1

Marinov M, Giubilei F, Gerhardt M, Özkan T, Stergiou E, Papadopol M, Cabecinha L (2013) Urban freight movement by rail. J Transp Lit 7:87–116

Motraghi A, Marinov MV (2012) Analysis of urban freight by rail using event based simulation. Simul Modell Pract Theor 25:73–89. ISSN 1569-190X. https://doi.org/10.1016/j.simpat.2012.02.009. http://www.sciencedirect.com/science/article/pii/S1569190X12000329

Niérat P (1997) Market area of rail-truck terminals: pertinence of the spatial theory. Transp Res Part A: Policy Pract 31(2):109–127. ISSN 09658564. https://doi.org/10.1016/s0965-8564(96)00015-8

Novaes AGN, Graciolli OD (1999) Designing multi-vehicle delivery tours in a gridcell format. Eur J Oper Res 119(3):613–634. ISSN 03772217. https://doi.org/10.1016/s0377-2217(98)00344-0

Novaes AGN, de Cursi JES, Graciolli OD (2000) A continuous approach to the design of physical distribution systems. Comput Oper Res 27(9):877–893. ISSN 03050548. https://doi.org/10.1016/s0305-0548(99)00063-5

Novaes AGN, Frazzon EM, Burin PJ (2009) Dynamic routing in over congested urban areas. In: Proceedings LDIC 2009, second international conference on dynamics in logistics. Bremen, pp 103–112

Pachl J, White T (2004) Analytical capacity management with blocking times. In: 83rd meeting of transportation research board. Washington DC, pp 1–14

Pillac V, Gendreau M, Guéret C, Medaglia AL (2013) A review of dynamic vehicle routing problems. Eur J Oper Res 225(1):1–11. ISSN 03772217. https://doi.org/10.1016/j.ejor.2012.08.015

SETRANS (2013a) Pesquisas de Origem e Destino – Parte 2: An´alise dos Resultados da Pesquisa com Passageiros de Autom´oveis no Cordon Line (in Portuguese). Technical report, Secretaria de Estado de Transportes, Rio de Janeiro RJ. http://www.rj.gov.br/web/setrans/

SETRANS (2013b) Minuta do Relatório 4 – Planejamento e Execução das Pesquisas: Parte 2: Tomo I – Pesquisanas Estações de Trem (in Portuguese). Technical report, Secretaria de Estado de Transportes, Rio de Janeiro RJ. http://www.rj.gov.br/web/setrans/

Sheffi Y (1985) Urban transportation networks: equilibrium analysis with mathematical programming methods. Prentice Hall, NJ

Singhania V, Marinov M (2017) An event-based simulation model for analysing the utilization levels of a railway line in urban area. Promet - Traffic Transp 29(5):521–528. https://doi.org/10.7307/ptt.v29i5.2306. https://traffic.fpz.hr/index.php/PROMTT/article/view/2306

Smilowitz KR, Daganzo CF (2007) Continuum approximation techniques for the design of integrated package distribution systems. Networks 50(3):183–196. ISSN 00283045. https://doi.org/10.1002/net.20189

Taniguchi E, Nemoto T (2008) Intermodal freight transport in urban areas in Japan. In: Konings R, Priemus H, Nijkamp P (eds) The future of intermodal freight transport, chapter 4. E. Elgarr, Cheltenham, UK, pp 58–65. ISBN 9781845422387. https://doi.org/10.4337/9781848441392.00008

Taniguchi E, Thompson RG (2004) Logistics systems for sustainable cities. Elsevier

van Woensel T, Cruz FRB (2009) A stochastic approach to traffic congestion costs. Comput Oper Res 36:1731–1739. https://doi.org/10.1016/j.cor.2008.04.008

Vuchic VR (2007) Urban transit systems and techonology. Wiley, Hoboken, NJ. ISBN 9780471758235

Simulation of Fire Dynamics and Firefighting System for a Full-Scale Passenger Rolling Stock

Ramy E. Shaltout and Mohamed A. Ismail

Abstract The secure travel and the safety of passengers are the utmost priority for transportation authorities all over the world. The fire safety in railway rolling stocks has gained significant importance in recent years. The study of fire dynamics including the fire growth and spreading, allows the development of fire protection techniques and passenger evacuation scenarios. Understanding fire development in train carriages is limited as few experimental investigations have been conducted on full-scale fire dynamic measurements of entire passenger coaches. This paper represents a comprehensive account of the computational fluid dynamics model used for the simulation of the fire dynamics in full scale rolling stock vehicle. Full-scale heat release rate (HRR) measurements were carried out for the entire vehicle as well as temperature distribution for various fire compartment scenarios. The model was tested and the simulation results were verified against those presented in experimental research developments in the literature.

Keywords Fire dynamics · Rail vehicles · CFD · Water-fire interaction · Firefighting

1 Introduction

Fire dynamics simulations are important because of their influence on the assessment of the thermal effects and threats on the design of evacuation procedures and plans, structural damage, and eventually the design of fire protection systems. Investigating fire development inside a rail vehicle aims to reduce the risk to passengers inside the vehicle as well as the users of the infrastructure. It permits the prediction of smoke

R. E. Shaltout (✉) · M. A. Ismail
Mechanical Power Engineering Department, Faculty of Engineering, Zagazig University, Zagazig, Egypt
e-mail: rashaltout@zu.edu.eg; rashaltout@gmail.com

R. E. Shaltout
Civil Engineering Program, German University in Cairo, New Cairo City, Egypt

© Springer Nature Switzerland AG 2020   209
M. Marinov and J. Piip (eds.), *Sustainable Rail Transport*, Lecture Notes in Mobility,
https://doi.org/10.1007/978-3-030-19519-9_7

and toxic hazards that might disturb the evacuation process of the passengers and the vehicle crew (GM/GN 2630 2013).

Numerical simulations using computational fluid dynamics (CFD) techniques have been used for the prediction of fire developments in train carriages (Zhang et al. 2017; Guillaume et al. 2014). The use of numerical simulation techniques in modelling fire development is a complex interdisciplinary topic. CFD models can be used in scenarios validated by experimental results or in qualitative studies (Guillaume et al. 2014; Li and Ingason 2016). However, by increasing the complexity in the railway tunnel and train designs, the use of CFD simulation is increased to predict the fire growth and smoke hazards for different simulation scenarios. Chiam (2005), presented a CFD model for predicting the HRR peak value for emergency tunnel ventilation systems. Chaim extended his study to include 13 fire scenarios using Fire Dynamics Simulator (FDS) to simulate the smoke spread and fire growth for a metro carriage in an underground subway tunnel. It was concluded that there is a significant need to enhance the modelling and simulation procedures used for the prediction of fire hazards for passenger trains, especially for scenarios where the carriage had stalled in a subway tunnel. Full-scale experimental studies were recommended by Chaim to validate the obtained simulation results.

Few experimental research activities have been conducted for investigating full-scale fire dynamic measurements of an entire passenger coach. This shortage in experimental investigations, has a significant impact on studying fire developments in train carriages and its impact on the vehicles' sustainability and reliability (Matsika 2018). The limitations in experimental research in such a field has influenced the understanding of fire behaviour, the economic impact, and implications for vehicle safety in terms of both manufacturing and material testing costs. It is worth mentioning that a number of tests were carried out in RSSB commissioned research projects such as T843 and T1012 (Tooley 2011; Rail Safety and Standards Board Ltd. 2014), for testing of the main elements of rolling stock interiors compared to both the British and European testing rules. These projects explored the weaknesses and strengths of the new standard EN45545 and identify the influence of the British industry in shifting from British standards such as RGS GM/RT2130 and BS 6853 (BS 6853 1999; GM/RT2130 2010).

An experimental study was carried out by White (White 2010), to investigate and analyse fire development in Australian metropolitan passenger trains. To estimate the Heat Release Rate (HRR) for a full-scale passenger train, the conservation of energy model was used based on experimental observation and measurements. The research also evaluated the existing methods used for fire estimations applied to passenger rolling stocks. Li et al. (Li et al. 2014; Li and Ingason 2016), experimentally investigated fire development in three series of train carriages. The fire tests were carried out for train carriages including a 1:10 model, a 1:3 model and a 1:1 full scale tunnel test. It was noticed that fire development was similar for the three test models. The HRR was also investigated for the three models of the train carriages. It was observed that the maximum HRR is dependent on the type of fuel, the train carriage configuration, material characteristics and the ventilation conditions. An overview of the experimental tests carried out on a full-scale passenger trains was presented by Li, Y. Z see (Li and Ingason 2016).

Fire performance characteristics of passenger train materials were investigated by Peacock et al. (2001). A cone Calorimeter tests were used to predict the smoke and HRR characteristics for the materials used in US passenger cars including the seats and interior panels. Afterwards, a full-scale test was carried out in a complete train car incorporating the tested components. The final results have been used in the fire safety requirements as well as providing a profitable way for designers and car builders to evaluate materials used in manufacturing passenger rail cars.

A comprehensive investigation of the dynamics of design fire in trains was presented in Hjohlman et al. (2009). A full-scale test was carried out on a complete passenger rail car to investigate the flashover design fire and its influence on the vehicle materials and design. Small scale tests were performed for the vehicle materials. The results of these tests were used as an input for the Fire dynamics Simulator (FDS) software. This was used to investigate the design fire for the tested sections. These do not vary greatly from the tested configuration for material selection and vehicle geometry. A good correlation was obtained between the simulation results using FDS and the full-scale experiment. It was noted that the results obtained from FDS model was significantly affected by the grid size.

The importance of designing fire protection systems and firefighting solutions has greatly increased for both existing and newly designed trains. Due to the safety measures presented by safety standard boards around the world and the European TSI (TSI LOC&PAS 2014), it is necessary to install, test and analyze the performance of fire protection systems. To compensate for the lack of existing standards and guidelines concerning the fire safety in passenger vehicles, a series of guidelines (ARGE) (ARGE Guideline 2012a, b, c) was introduced to evaluate the system performance and to define the necessary measures for the firefighting systems.

The main objective of the presented work in this paper, is to study the development of fire dynamics in passenger trains in order to establish an efficient fire protection system for such important assets. A methodology was proposed in this research that incorporates a series of sprinklers distributed through the vehicle ceiling. 83% reduction in HRR was achieved when using water-spray with the fire, which gives more time to allow passengers safe evacuation.

2 Proposed Methodology

The computational models presented in this paper were built in the Fire Dynamic Simulation (FDS) package. Existing three different models used to simulate the solid fuel burning in the FDS software. The user should define the type of models used in the analysis according to the input parameters including the ignition source, material characteristics and the fire load. The first model used in the analysis is (HRRPUA). It simulates a gas burner with a determined Heat Release Rate Per Unite Area. The use of HRRPUA in FDS was validated by comparing the results with experimental tests

(Li et al. 2014; Ma and Quintiere 2003). The second model is the simple ignition model. In this model, when the material surface temperature reaches the ignition temperature then the ignition temperature is assigned to a combustible material. The third model is the Kinetic pyrolysis model (Li et al. 2014). In this model, the reaction is described by means of the kinetic parameters defined for the combustion of each material. The simple ignition model and the kinetic pyrolysis model are still under development and needs more verifications in case they are used to predict the fire development in different scenarios (Li et al. 2014). In this paper, HRRPUA is used in the analysis. The prescribed fire is defined throughout burning wood cribs in the right corner of the vehicle.

3 Computational Model Description

Modelling and simulation of fire development in train carriages can be used as a suitable alternative option for making possible design changes as well as investigating all parameters that could influence the fire dynamic behaviour. Small scale laboratory tests have been used by many authors as an input data to the CFD computational tools to model the fire development in passenger rolling stocks (White 2010; Li et al. 2014; Li and Ingason 2016; Lönnermark et al. 2017). These tests included for example, the Cone Calorimeter and Thermo-Gravimetric Analysis (TGA) tests. In this paper, Fire Dynamic Simulation (FDS) software has been used in modelling the fire development and smoke spread for a 1/3 train mock-up model and for a full-scale carriage respectively.

FDS software (McGrattan et al. 2013, 2015), is still widely used for simulating the design fire in train compartments which do not vary considerably from the tested configurations. The simplifications and assumptions in the FDS model are in the input parameters, especially for the combustion model and the combustible glazing and materials, which might lead to a lake of realistic design fires prediction (Guillaume et al. 2014; White 2010; Li and Ingason 2016). The obtained results from the FDS model built in this paper for the 1/3 train carriage, were compared to those obtained by Lönnermark et al. (2017), for the validation purposes of the model and the methodology used in the analysis. The model is then extended to include a full scale carriage.

3.1 Geometry Description

3.1.1 1/3 Train Carriage Mock-up

The preliminary analysis in this paper was carried out based on a model described by Lönnermark et al. (2017). A model with 6 m × 3 m × 2.4 m (L × W × H) dimension was defined for a 1/3 train carriage as shown by Fig. 1. To simulate the

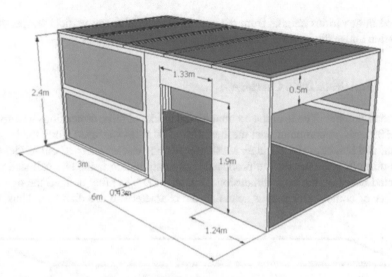

Fig. 1 1/3 train mock-up used in the experiment (Lönnermark et al. 2017)

smoke layer in the carriage, an edge was created with a height of 0.5 m. The fire source was placed at the corner of the model. The inner layers of the carriage ceiling, walls and linings were initially covered by noncombustible boards (Promatect H). For more details see (Lönnermark et al. 2017).

3.1.2   Full-Scale Train Carriage

In this paper, fire development in a full train carriage was investigated. The dimensions of the entire train carriage were 18 m × 3 m × 2.4 m in x-, y-, and z- direction respectively. The governing equations were approximated on a three-dimensional mesh with cubic elements (Fig. 2).

3.2   CFD Model and Parameters

3.2.1   1/3 Train Carriage Mock-up

K-type thermocouples have been used for the gas temperature measurements using 0.5 mm welded junction. In some positions, an 0.25 mm K-type thermocouple was placed to estimate the effect of radiation on the temperature measurement. The locations of the thermocouple are shown in Fig. 5. Thermocouples were located at different heights from vehicle ceilings. Some of them were placed 0.29 m below the ceiling while two others were sited at 0.05 m below the ceiling. A thermocouple tree was placed in at various heights from the ceiling to determine the vertical temperature variation above the ignition source position (Fig. 3).

3.2.2 Full Scale Train Carriage

The total number of cells of the computational model of three dimensions was about 129,600 cells. A computational mesh of $180 \times 30 \times 24$ was used along the length, width, and height of the carriage. Mesh dependency test was carried out to identify the optimum grid size for the proposed computational domain. The cell size was selected based on the non- dimensional quantity $(D^*/\partial x)$ that defined the optimal number of computational cells spanning the characteristic fire diameter. Thus, the

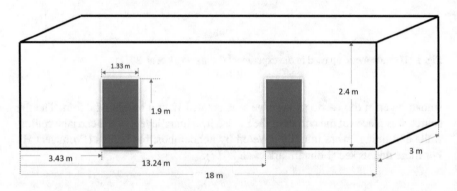

Fig. 2 Schematic diagram of the full train carriage used in the simulation

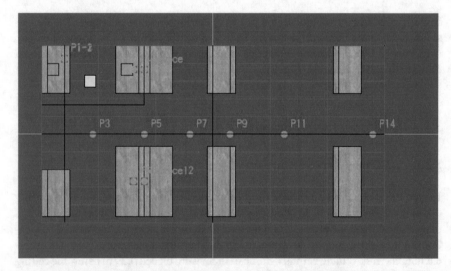

Fig. 3 Thermocouples distribution for FDS model of the 1/3 train carriage mock-up

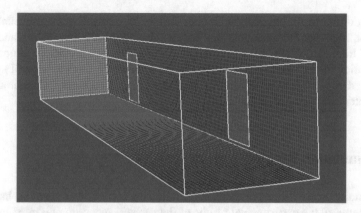

Fig. 4 Computational model of full-scale carriage

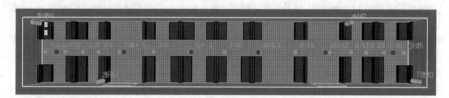

Fig. 5 Thermocouples distribution along the full-scale train carriage

computational cells were almost cubic with dimensions of $0.1 \times 0.1 \times 0.1 \ m^3$ as shown by Fig. 4.

Following the same procedure followed in the mock-up model, thermocouples were distributed along the full-scale carriage.

To determine the temperature distribution in the full-scale train carriage, 25 thermocouples were distributed according to the dimensions given in Table 1. 5 thermocouples trees were defined in the model. Each tree contains 5 thermocouples numbered as shown by Table 1 (Fig. 5).

Table 1 Thermocouple trees position and distribution along the full-scale carriage model

Thermocouples tree #	Thermocouple	Thermocouple tree position	
		X	Y
1	$T_{1-1}:T_{1-5}$	0.3	2.7
2	$T_{2-1}:T_{2-5}$	3.2	0.3
3	$T_{3-1}:T_{3-5}$	9	1.5
4	$T_{4-1}:T_{4-5}$	14.8	0.7
5	$T_{5-1}:T_{5-5}$	17.3	0.3

216   R. E. Shaltout and M. A. Ismail

In addition to the thermocouple trees defined in Table 1, a thermocouple tree contains 15 K-type thermocouple T_{6-1}:T_{6-15} was installed at the horizontal plane along the full train at a height of 0.5 m from the ceiling. This tree provides a temperature distribution for the smoke layer in the carriage. Another thermocouple tree T_{7-1}:T_{7-15} was fitted on a height of 0.29 below the carriage ceiling. A summary of the simulation parameters are included in Table 2.

4 Simulation Results

The simulation results are presented in the following section. A comparison has been made between the experimental results of the carriage mock-up presented by Lönnermark et al. 2017) and those obtained by the developed simulation model in this paper. The main objective of the comparison is to validate the computational model and the applied fire development criteria proposed in this paper. Subsequently, simulation results of a full-scale train carriage will be presented. Various fire development scenarios were also investigated. The HRRPUA used in the analysis of the full car varied from 400 to 8800 [kW/m^2] to simulate several fire ignition setups. Finally, simulation results of a 1/3 train mock-up model and full-scale train model incorporating a series of sprinklers, as a firefighting mechanism, were illustrated. A fire source in the corner of the vehicle was proposed with various heat release rates per unit area (HRRPUA) varying from 400 to 8,800 kW/m^2. The lower limit simulates a small fire source, while the upper limit of the HRRPUA simulates a very big fire source or terroristic attack. At this case, the total HRR reached 11,800 kW after 70 s and the maximum temperature reached to 1300 °C after 40 s. The design of the water-spray firefighting system was proposed in the presented paper. It was found that the maximum HRR reduces from 11,800 to 2000 kW (83% reduction) when using water spray with the fire and it was delayed 30 s more, which gives more time to allow passengers evacuation.

4.1 Test Characteristics

The experimental work presented in Lönnermark et al. (2017) included a total number of six tests. Three tests were selected for the validation purposes of the obtained results from the developed computational model in this paper. In Table 3, the maximum heat release rate and time to maximum for each of the selected tests are used in the comparison.

As it can be noticed in from the simulation results, the time to reach the maximum HRR in the simulation is quite similar to the time obtained from the experimental tests with a permissible computational variation.

Table 2 Simulation parameters for the 1/3rd train mock-up model and the complete carriage model

Model	Model dimension (L × W × H) m	Fire load	Fire source location	Variation of maximum HRRPA	Thermometer number-type	Simulation scenarios
1/3 train mock-up	6 × 3 × 2.4	Wood cribs	Left corner	360–3500 kW/m²	14 K-type	Without sprinklers
						With sprinklers
Full train carriage	18 × 3 × 2.4	– Wood cribs – Small bag	Left corner	400–8800 kW/m²	25 K-type	Without sprinklers
						With sprinklers

4.2 Mock-up Model Simulation Results

In this section, the simulation results for the 1/3 mock-up carriage is compared to the experimental data for the model validation. Three tests were compared but only test#1 (Lönnermark et al. 2017) will be discussed here. As shown in Fig. 6, the heat release rate peak for both simulation and experimental cases are identical. The heat release rate reaches a maximum of 355 kW after a time of 6:01 min for the simulation results. This was compared to the experimental data which has a maximum heat release rate of 360 kW after 6:12 min as shown in Table 3. The error in the maximum heat release rate between the simulation results and the experimental data was found to be 1.4% which is acceptable for model validation (Anderson and Wendt 1995). This small error might come from not accounting for all the actual materials of seats used in the real test. To get an overview of the quantity of the carriage material burned, the mass loss rate is plotted against the heat release rate for Test#1 as shown in Fig. 7. It is clear from that figure that mass loss rate is following the same trend as the heat release rate curve. The mass loss rate reaches a maximum of 0.92 kg/s at time of 6:00 min which is slightly before the maximum heat release rate. These phenomena are common in natural fires because mass is lost first in a fire followed by heat release (Drysdale 2011).

The temperature along the vertical plane directly above the fire source was measured by thermocouple tree#1, which contains 5 thermocouples. T_{1-1} is the nearest to the fire source and T_{1-5} is the furthest from the fire source and nearest to the ceiling. As illustrated in Fig. 8, all the thermocouples have the same trend which have peaks after 6:00 min from starting fire. This delay in temperature peaks comes from the time needed to get maximum heat release rate and transfer of this heat by diffusion and radiation. The maximum temperature decreases in the upward direction and going further from the fire source. Temperatures in tree#1 varies from T_{1-1} = 610 °C to T_{1-5} = 279 °C. At the time of maximum heat release rate (6:01 min), the temperature varies horizontally away from the fire source. At a height of 0.29 m below the ceiling, the temperature varies from 264 °C at 0.7 m horizontally from the fire source to 173 °C at the end of the carriage mock-up. It should be noted here that during test#1 no flashover occurs, and the fire does not move to the seat facing the fire source.

Table 3 Maximum HRR and time to reach it for simulation tests versus their corresponding experimental data from Lönnermark et al. (2017)

Test #	Maximum HRR (kW)		Time to max. HRR (min)	
	Exp.	Sim.	Exp.	Sim.
1	360	365	6:12	6:04
3	640	590	5:30	5:20
6	3500	3450	7:00	7:10

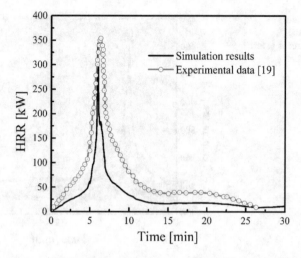

Fig. 6 Comparison between the obtained HRR from the FDS simulation results of and the experimental results for test #1 documented in Lönnermark et al. (2017)

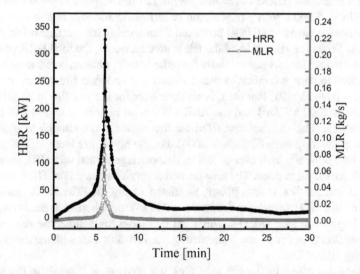

Fig. 7 Simulation results of heat release rate (HRR) and mass loss rate (MLR) for test#1

4.3 Full-Scale Carriage Simulation Results

After verification of the model against the 1/3 carriage mock-up, the model was expanded to include the full-scale carriage as shown in Fig. 5. In this section, the fire source with various heat release rate per unit area (HRRPUA) will be used. The position of the fire source is fixed at the left corner of the carriage above the seat that lies under the thermocouple tree #1, Fig. 5. The fire source will vary from

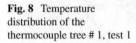

Fig. 8 Temperature
distribution of the
thermocouple tree # 1, test 1

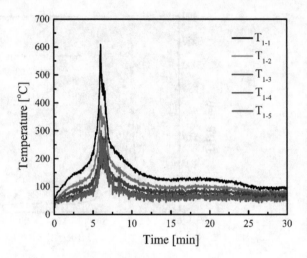

burning a small bag (HRRPUA = 400 kW/m^2) to simulating a larger source of fire
(HRRPUA = 8,800 kW/m^2) (e.g. Arson or terroristic attacks). Figure 9 illustrates
the total heat release rates (HRR) produced from burning the carriage using various
fire loads. When the HRRPUA of the fire source increases, the total HRR produced
from the carriage increases, especially from the low fire powers, because more seats
are exposed to burn according to heat radiated from a larger fire source. It can be
concluded from Fig. 9a, that there is no flash over for the low fire loads (HRRPU
from 400 to 1000 kW/m^2) and the HRR did not go high beyond 800 kW for the
whole vehicle. The radiated heat from the fire source is not enough to ignite any
of the neighbouring seats (Quintiere 2006). For the higher fire loads (HRRPU from
2000 to 8800 kW/m^2) flash over occurs and the carriage burned with HRR exceeding
6000 kW for the three cases. The time needed to reach the maximum HRR decreases
by increasing the fire source power, as shown in Fig. 9b. The HRR reached its
maximum after 800 s when the fire load is 2000 kW/m^2 whereas, it needs only 70 s
for the largest fire load (8800 kW/m^2). At this large fire source, the full carriage
takes only 90 s to be fully burned simulating a terroristic attack with maximum HRR
exceeding 10,000 kW.

To show the effect of the fire source on the carriage, it is better to discuss the
temperature behavior along the vehicle. Figure 10 illustrates the vertical variation of
temperature at various positions for the lowest fire load (400 kW/m^2). The vertical
thermocouple Tree# 1, located directly above the fire source, shows that temperature
decreases going higher above the fire source with large differences in temperature
between different positions. The temperature has a value of 800 °C above the fire
source and decreases to 540 °C at 0.29 m below the ceiling, Fig. 10a. Going away
from the fire source at the location of thermocouple Tree# 5, the vertical variation
of temperature is very small which indicates the homogenous distribution of temperature at this position as shown in Fig. 10b. Figure 11 illustrates the horizontal
distribution of temperature along different planes, at 0.29 and 1.0 m from the ceiling,

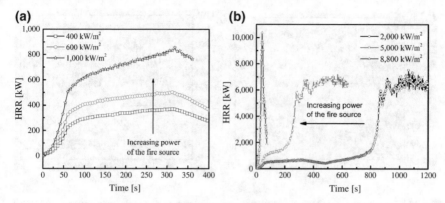

Fig. 9 Comparison between the obtained HRR from the FDS simulation results of a full-scale carriage

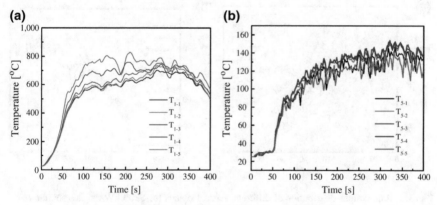

Fig. 10 Temperature distribution at vertical planes for 400 kW/m² fire source; **a** tree 1 above the fire source and **b** tree 5 at the far end of the carriage and opposite to fire source

for the case with smallest fire load (400 kW/m²). For the two planes, temperatures at T_{1-3} and T_{1-5} are much higher than all other thermocouples, evidence of no burning of the carriage. The heat from the fire source does not raise the temperature in positions away from it, indicating no flash over.

Figure 12 illustrates the vertical temperature distribution at different positions for the case with maximum fire load (8800 kW/m²) of the full carriage. In this case, all the thermocouples in Tree# 1 read almost the same temperature with a maximum value of 1150 °C at the time of maximum HRR. Very high HRR due to rapid burning of the full carriage, indicates the temperature is homogenous vertically but has lower values going away from the fire source. For the thermocouple Tree# 5, the temperature rise is delayed more than Tree# 1 because of the time needed for the HRR to reach the other end of the vehicle.

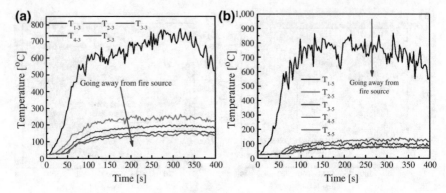

Fig. 11 Temperature distribution at different horizontal planes across the carriage for 400 kW/m^2 fire source at heights of 0.29 and 1.0 m below the ceiling for **a** and **b**, respectively

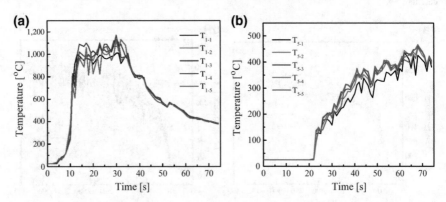

Fig. 12 Temperature distribution at different vertical planes for 8800 kW/m^2 fire source; tree 1 above the fire source and tree 5 at the far-end of the carriage and opposite to fire source

Figure 13 shows the horizontal temperature distribution across the full carriage for the highest fire load (8800 kW/m^2). In the first 30 s, the nearest thermocouples to the fire source (T_{1-3} and T_{1-5}) have a higher reading than other ones, but after 30 s the temperature is almost homogenous as the heat is transferred from the fire source to the different positions in the carriage.

4.4 Simulations Incorporating Firefighting System

To face the fire generated in the carriage, automatic water sprinkler systems are proposed. Based on fire standards (Lataille 2002; Puchovsky 1999), the Pendent sprinkler type with activation temperature of 79 °C and particle velocity of 5 m/s is suggested. Three automatic sprinklers are fixed in the 1/3 mock-up at a height

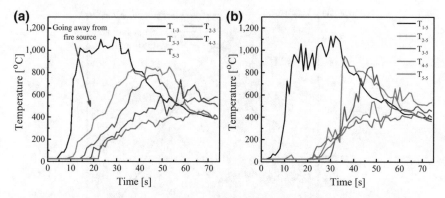

Fig. 13 Temperature distribution at different horizontal planes across the carriage for 8800 kW/m^2 fire source

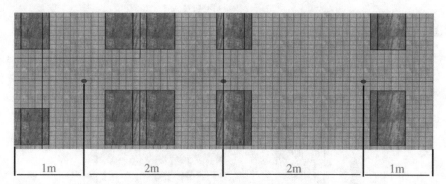

Fig. 14 Sprinklers distribution along the 1/3 mock-up

of 2.35 m from the carriage floor with water flow rate of 200 L/min, as shown in Fig. 14. To study the effect of the water sprinkler on fire HRR and temperature inside the 1/3 mock-up, test #6 documented in Lönnermark et al. (2017) is simulated using sprinklers and compared to the original case without using sprinklers. It is clear from Fig. 15 that water sprinklers reduce the maximum HRR by 74.8% from 3500 to 880 kW. The temperatures also decrease from 950 to 90 °C as shown in Fig. 16. The huge reduction in HRR and temperature provides positive advantages for evacuating passengers inside the vehicle in case of any fire accident.

The temperature distribution along the horizontal plane (0.29 m below the ceiling) across the mock-up is shown in Fig. 17. Without sprinklers, the temperature has the value of 950 °C and decreases away from the fire source with a value of 550 °C at the end of the mock-up (5 m away from fire source). After using water sprinklers, the temperature reduces to a homogenous temperature of 90 °C along the mock-up.

Figure 18 shows the effect of water flow rate of the automatic sprinkler on the HRR produced from the 1/3 mock-up. Without any water sprinklers, the maximum HRR was 3500 kW. When water was used with a flow rate of 100 L/min, the reduction

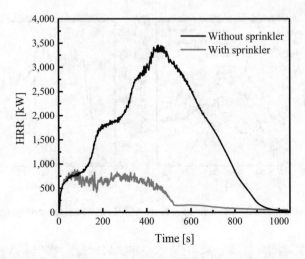

Fig. 15 Comparison between the obtained HRR from the FDS simulation results with and without sprinklers for test# 6 documented in Lönnermark et al. (2017)

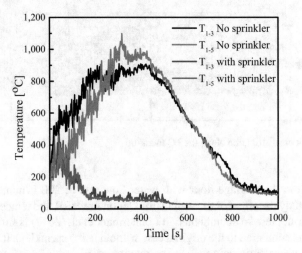

Fig. 16 Some temperature distributions in the 1/3 mock-up with and without sprinklers

in HRR was 69% and going further to 74.8% with flow rate of 200 L/min. For the maximum water flow rate of 300 L/min, the reduction in HRR reached 77%. It is clear from this figure that increasing the water flow rate further after 200 L/min will reduce the maximum HRR with small percentages, so it is recommended not to go beyond 200 L/min of water flow rate.

The sprinkler model is expanded to cover the full-scale carriage, as shown in Fig. 19. The first sprinkler is fixed at a position of 1.0 m from the start of the carriage with 2.0 m between each of the two sprinklers. Like the 1/3 mock-up, 9 sprinklers are fixed at a height of 2.35 m from the carriage floor with water flow rate of 200 L/min.

Fig. 17 Temperature distribution along the horizontal plane (0.29 m below the ceiling) across the 1/3 mock-up with and without sprinklers

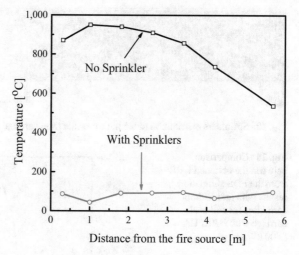

Fig. 18 Effect of sprinkler's water flow rate on the maximum heat release rate produced from fire in mock-up

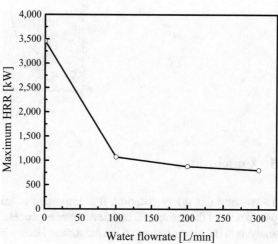

The highest HRR case (8800 kW/m²) is modeled with the sprinkler in the full-scale carriage and the comparison with sprinkler is shown in Fig. 20. The water sprinklers reduce the maximum HRR from 10,366 to 2013 kW with a reduction percentage of 80.6% which saves the carriage from burning and gives enough time for passengers to escape in case of fire. It can be concluded that the water sprinkler system is very efficient in suppressing fire inside the train coach.

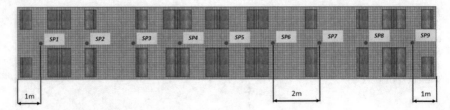

Fig. 19 Sprinklers distribution along the full-scale train carriage

Fig. 20 Comparison between the obtained HRR from the FDS simulation results with and without sprinklers for full-scale carriage with HRRPUA = 8800 kW/m^2

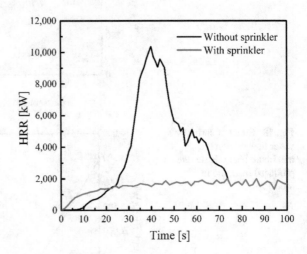

5 Conclusions

To evaluate fire safety criterion, it is important to initially identify the controlling parameters of the fire spread and development process. A crucial part in fire dynamic analysis is the measurement of the maximum HRR which will describes how fast the process is in reaching a fully developed fire in a train compartment. In this paper, a comparison has been made between the experimental results for a small-scale train coach and those obtained from the CFD model proposed in the presented work. The main objective of the comparison is to check the validity of the obtained results and to assess the reliability of the model to be used in the analysis of fire dynamics in train cars with different fire scenarios.

By varying the fire loads HRRPU from 400 to 1000 kW/m^2, no flash over was detected as the radiated heat from the fire source was insufficient to ignite any of the neighbouring seats from the fire source located on the left corner seat. But, in case the fire loads increased from 2000 to 8800 kW/m^2, the HRR in that case exceeded the 6000 kW and flash over was detected.

For the purposes of protecting rail assets, this paper investigated a fire protection system. It was found that the maximum HRR was achieved by 80.6%, by incorporating a water fire interaction mechanism using distributed sprinklers in the full train.

In this way the system delayed the complete burning of the full carriage and gives enough time for passengers to escape in case of fire accidents.

From the quality of the results obtained in the analysis of the fire development in a full train carriage, it can be concluded that the simulation model presented is reliable and efficient. It can be used in the fire dynamic analysis of different scenarios. Further work is still needed to study the impact of changing the fire source position in the train carriage and its influence on the fire and smoke spread as well as flashover in the whole carriage. In addition, investigating the effect of changing the sprinkler distribution and water mist flow on the fire and smoke development speed is required.

References

Anderson JD, Wendt J (1995) Computational fluid dynamics, vol 206. McGraw-Hill, New York
ARGE Guideline—Part 1 (2012a) Fire detection in rolling stock
ARGE Guideline—Part 2 (2012b) Fire fighting in rolling Stock
ARGE Guideline—Part 3 (2012c) System functionality of fire detection and fire-fighting systems in rolling stock
BS 6853:1999. Code of practice for fire precautions in the design and construction of passenger carrying trains
Chiam BH (2005) Numerical simulation of a metro train fire. Doctoral thesis, Department of Civil Engineering, University of Canterbury, New Zealand
Drysdale D (2011) An introduction to fire dynamics. Wiley, New York
GM/GN2630 Guidance on Rail Vehicle Fire Safety, Issue One (2013) Rail industry guidance, note for GM/RT2130. Issue Four. Part 2
GM/RT2130, Vehicle fire safety and evacuation, Railway group Standards, Rail Safety and Standard Board Limited, 2010
Guillaume E, Camillo A, Rogaume T (2014) Application and limitations of a method based on pyrolysis models to simulate railway rolling stock fire scenarios. Fire Technol 50(2):317–348
Hjohlman M, Försth M, Axelsson J (2009) Design fire for a train compartment
Lataille J (2002) Fire protection engineering in building design. Elsevier, Amsterdam
Li YZ, Ingason H (2016) A new methodology of design fires for train carriages based on exponential curve method. Fire Technol 52(5):1449–1464
Li YZ, Ingason H, Lonnermark A (2014) Fire development in different scales of train carriages. Fire Safety Sci 11:302–315
Lönnermark A, Ingason H, Li YZ, Kumm M (2017) Fire development in a 1/3 train carriage mock-up. Fire Saf J 91:432–440
Ma TG, Quintiere JG (2003) Numerical simulation of axi-symmetric fire plumes: accuracy and limitations. Fire Saf J 38(5):467–492
Matsika E (2018) Interior train design of commuter trains: standing seats, and consideration for persons with reduced mobility. In: Marinov M (ed) Proceedings of RailNewcastle Talks 2016, sustainable rail transport. Springer, Cham, pp 59–75
McGrattan K, Hostikka S, McDermott R, Floyd J, Weinschenk C, Overholt K (2013) Fire dynamics simulator technical reference guide volume 1: mathematical model (version 6). National Institute of Standards and Technology, USA
McGrattan K, Hostikka S, McDermott R, Floyd J, Weinschenk C, Overholt K (2015) Fire dynamics simulator user's guide (version 6). National Institute of Standards and Technology, USA
Peacock RD, Bukowski RW, Reneke PA, Averill JD, Markos SH (2001) Development of a fire hazard assessment method to evaluate the fire safety of passenger trains. In: 7th international conference and exhibition

Puchovsky MT (ed) (1999) Automatic sprinkler systems handbook. National Fire Protection Association (NFPA)

Quintiere JG (2006) Fundamentals of fire phenomena. Wiley, Chichester

Rail Safety and Standards Board Ltd. (2014) Developing a good practice guide for managing personal security on-board trains

Tooley D (2011) A comparison of New European fire standards with UK standards and the impact on UK vehicle design. Proc Inst Mech Eng Part F J Rail Rapid Transit 225(4):403–416

TSI LOC&PAS—Technical Specification for Interoperability for Locomotives and Passenger Cars, in force by Commission Regulation (EU) 1302/2014

White N (2010) Fire development in passenger trains. Doctoral dissertation, Victoria University

Zhang S, Yang H, Yao Y, Zhu K, Zhou Y, Shi L, Cheng X (2017) Numerical investigation of back-layering length and critical velocity in curved subway tunnels with different turning radius. Fire Technol 53(5):1765–1793

Novel Energy Harvesting Solutions for Powering Trackside Electronic Equipment

Cristian Ulianov, Zdeněk Hadaš, Paul Hyde and Jan Smilek

Abstract Recent developments in different areas have enabled the improvement and development of new energy harvesting technologies that could potentially be successfully employed for various railway applications. The state of development of energy harvesting solutions potentially suitable for integration in the railway environment to power trackside equipment has been reviewed and assessed. The general harvesting capacities and characteristics of potential energy harvesting technologies have been discussed, along with the general power usage requirements and characteristics of common types of trackside equipment. Conclusions have been drawn about the most suitable energy harvesting technologies, or combination of technologies to be incorporated into a combined energy harvesting and storage power supply for different trackside equipment.

Keywords Energy harvesting · Railway systems · Signalling · Trackside electronics

1 Introduction

Trackside equipment used to control and monitor the movement of train and monitor the railway infrastructure are currently generally powered through cables from the electricity grid. This has significant impacts on overall infrastructure reliability, availability, maintainability and safety (RAMS), as well as on its life cycle costs (LCC). Developments in energy harvesting technologies, and low-power and cost-efficient sensors and communication equipment

C. Ulianov (✉) · P. Hyde
NewRail Centre for Railway Research, Newcastle University, Newcastle upon Tyne, UK
e-mail: cristian.ulianov@ncl.ac.uk

Z. Hadaš · J. Smilek
Institute of Solid Mechanics, Mechatronics and Biomechanics, Brno University of Technology, Brno, Czech Republic

© Springer Nature Switzerland AG 2020
M. Marinov and J. Piip (eds.), *Sustainable Rail Transport*, Lecture Notes in Mobility,
https://doi.org/10.1007/978-3-030-19519-9_8

, have made the development of self-powered energy harvesting trackside systems a realistic possibility. The development of energy harvesting technologies that could be easily integrated into the track system is a potentially viable alternative powering solution for trackside equipment, particularly if combined with more power efficient trackside equipment.

Such energy harvesting solutions would allow minimising the use of cables and provide infrastructure managers with alternative technologies that could be used for the next generation of low-power trackside electronic equipment.

In this paper, the energy harvesting capabilities and technology characteristics have been considered, with respect to the energy requirements of common trackside applications. An overview of combinations of energy harvesting technologies and applications that could be potentially viable is presented.

A commonly used definition of energy harvesting technologies (Mateu and Moll 2005) is:

> An energy harvesting device generates electric energy from its surroundings using some energy conversion method. Therefore, the energy harvesting devices here considered do not consume any fuel or substance. On the other hand, as the environment energy levels are very low (at least for today's electronic devices requirements).

The energy harvesting concept is based on converting some type of available ambient energy into usable electrical energy using a dedicated device (energy harvester). The amount of available power and the ease or difficulty of its extraction and conversion are crucial limiting factors for independent devices, especially those of small size. Most of the current remote electronic applications rely on the battery or mains supply as a primary source of power. Energy harvesting, if implemented, serves mostly as a secondary power source meant to extend the service life until the next battery replacement or recharge. However, with the ongoing miniaturisation of the electronic devices, their increasing power efficiency and decreasing power consumption, the energy harvesters could serve as the primary power source for some low power applications. Larger energy harvesting and storage installations could be also considered for the primary power source of some trackside applications with moderate or high power requirements. However, the main determining factors for such large power supplies are likely to be life cycle cost and reliability rather than their technical viability.

2 General Considerations on Energy Harvesting Technologies

2.1 Background of Trackside Energy Harvesting Technologies

For nearly 20 years energy harvesting has been investigated as a possible source of power for wireless applications, which would be for one reason or another difficult

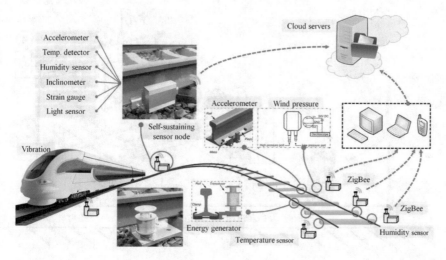

Fig. 1 Illustration of proposed solution for intelligent monitoring of underground railway (Gao et al. 2017a, b)

to connect to the power grid. These applications currently include mainly wireless sensor nodes for structural or health monitoring systems (Aktakka and Najafi 2014), in aerospace (Hadas et al. 2014) or transportation (Yoon et al. 2013). Figure 1 shows a solution proposed for intelligent monitoring of railways, which is based on energy harvesting technologies (Gao et al. 2017a, b).

Energy harvesting solutions also cover common renewable energy technologies, which are integrated in an original object. In case of trackside energy harvesting applications, it could be, e.g., a photovoltaic cover on trackside objects, trackside wind turbines, sleepers with integrated photovoltaic panel or sleeper with integrated wind turbine, etc. Wind turbines and solar panels are commonly used as autonomous sources of energy for various remote applications. These technologies can be exploited and integrated for the trackside power source solutions and turbines or solar panels provide alternative source of energy for trackside objects.

Kinetic energy in a form of mechanical deformation, vibrations, shocks, and thermal energy in a form of waste heat sources are exploitable inputs for autonomous power sources in several engineering applications related to energy harvesting. Waste heat sources are not readily available in the trackside environment. However, they could be useful for on-board solution (e.g., waste heat caused by friction). On the other hand, a passing train provides a non-negligible source of input mechanical energy, which could be converted into electricity by various transducer setups. Although a typical train does impart a large amount of mechanical energy into the infrastructure, the energy is distributed and dissipated over a large area, therefore, capturing this energy efficiently in a relatively small device is a complex problem. The amount of harvested energy from vibrations is usually very low, and the output electrical power

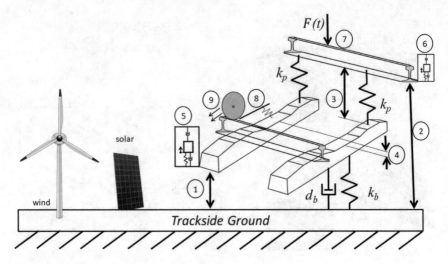

Fig. 2 Potential trackside energy harvesting technologies

has to be predicted and compared with the power requirements of intended ultra-low power applications.

Figure 2 shows potential application and integration of trackside energy harvesting technologies, including:

- Commercial solar and wind power products
- Displacement, strain and deformation energy harvesting—type *1, 2, 3, 4, 7, 8*
- Vibration energy harvesting—type *5, 6*
- Change of magnetic field by passing wheel—*9*
- Thermal gradient between in trackside environment (if available).

2.2 Physical Principles of Energy Harvesting

The scope of energy harvesting is to convert energy from one form to another, so that it could be used to power electronic devices. In general, energy harvesting relies on both ambient and external sources. Ambient sources such as radio frequency (RF), solar, thermal, wind, etc. are accessible within the environment, without any external energy supply. External sources (e.g., mechanical) are those that emit energy to the environment, with the intent for this energy to be harvested by specifically designed devices.

The underlying physical and technological principles, which can be applied to deriving power from various energy sources for the purpose of harvesting energy, are briefly summarised in this section.

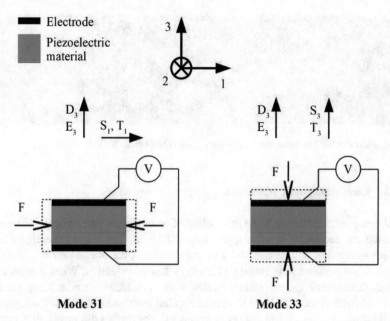

Fig. 3 Operation modes of piezoelectric material for energy harvesting

2.2.1 Piezoelectric Conversion

Piezoelectric materials have the property of converting a mechanical stress or strain applied to them into a change of electric field or electric displacement within the material and vice versa. Conversion from mechanical to electrical energy is called the direct piezoelectric effect, and the conversion from electrical into mechanical domain is known as reverse piezoelectric effect (Batra and Alomari 2017). The piezoelectric phenomenon is based on the fundamental structure of a crystalline network—certain crystalline structures have a charge balance with polarization, which must be oriented in one direction to produce piezoelectric behaviour of the material.

Two operation modes of piezoelectric materials are important for energy harvesting (Ambrosio et al. 2011), i.e.: mode 31 and mode 33, which are depicted in Fig. 3. In these modes, the external force is applied only in one direction, which is the most common case in energy harvesting devices and electrical potential is observed on electrodes.

A cantilever design of energy harvester, which uses the mode 31 of piezoelectric effect to harvest energy for bridge monitoring purposes was developed by Cahill et al. (2018) (Fig. 4).

Fig. 4 Piezoelectric cantilever for energy harvesting (Cahill et al. 2018)

2.2.2 Electromagnetic Induction

Electromagnetic induction is the production of an electromotive force (EMF) across an electrical conductor in a changing magnetic field. The EMF generated due to relative movement of a circuit and a magnetic field is the phenomenon underlying electrical generators and is based on Faraday's law of induction. When a permanent magnet or magnetic circuit is moved relative to a conductor, or vice versa, an electromotive force is created. If the wire is connected through an electrical load, current will flow, and thus electrical energy is generated, converting the mechanical energy of motion to electrical energy.

2.2.3 Electrostatic Conversion

The principle of electrostatic energy conversion lies in exploiting a capacitor with variable capacitance value. The two electrodes of the capacitor, separated by air, vacuum or any dielectric material, move with respect to each other due to mechanical excitation. That leads to a change either in the active surface of the electrodes, or their distance from each other, causing a variation in the capacitance (Boisseau et al. 2012). Energy harvesters based on electrostatic conversion are related to Micro-Electro-Mechanical Systems (MEMS) technologies and provide very low output power.

2.2.4 Magnetostriction

A characteristic property of magnetostrictive materials is that a mechanical strain will occur if they are subjected to a magnetic field in addition to strain originated from pure applied stresses, Fig. 5. Also, their magnetisation changes due to changes in applied mechanical stresses in addition to the changes caused by the changes of the applied magnetic field. A coil is integrated for energy harvesting operation (Kaleta et al. 2014). Commonly used magnetostrictive materials include Terfenol-D alloy (Fan and Yamamoto 2015), Galfenol (Berbyuk 2013) and Metglas.

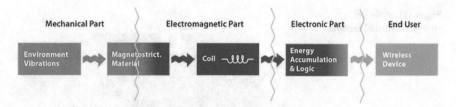

Fig. 5 Magnetostriction energy harvester as combination of smart material and electromagnetic induction

2.2.5 Triboelectric Effect

Triboelectric nanogenerators (TENG) are based on two principles such as triboelectric effect and electrostatic induction. The theory of this type of energy harvesters is described by many researches (e.g. Jiang et al. 2016), and follows from the structure of TENG device. TENG has shown advantages such as high output voltages, high energy-conversion efficiency, abundant choices of materials, scalability and flexibility (Zi et al. 2015). There is a potential to use as a power-generator floor or bed (Zhang et al. 2015).

2.2.6 Photovoltaic Effect

The photovoltaic effect occurs when photons are absorbed at a junction between two dissimilar materials (a heterojunction), inducing a voltage. Materials used for fabricating such heterojunctions are generally semiconductors, which are responsive to light of various wavelengths. A typical photovoltaic device mainly consists of a large area semiconductor p-n junction.

2.2.7 Thermoelectric Conversion

Thermoelectric generators consist of a thermoelectric module, a heat source (hot side) and a heat sink (cold side). The thermoelectric module utilises the Seebeck Effect to convert a temperature difference between each side of the device to an electromotive force. A schematic representation of a thermoelectric module is shown in Fig. 6. This phenomenon is based on diffusion of electrons through an interface between two different materials—usually semiconductors; the diffusion is achieved by applying heating at the junction of the materials, which make a thermocouple.

Fig. 6 Thermoelectric
energy harvesting module

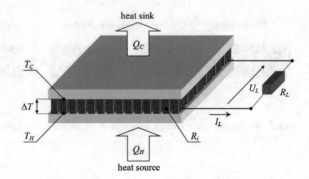

3   Implementation of Energy Harvesting Technologies in Trackside Applications

Development of effective implementations of energy harvesting for trackside applications depends on three main categories of factors. These are: the energy consumption of the device or devices being powered, the energy harvesting characteristics and capacity of the energy harvester, and the characteristics of the trackside environment.

3.1   Potential Trackside Applications of Energy Harvesting Technologies

The scope of trackside energy harvesting applications is to power trackside equipment locally and avoid the need for power cables from the electricity grid to the equipment. Associated with this, is the replacement of communications cables with radio communications, also powered by energy harvesting. The objective of reducing or eliminating trackside cables is to reduce installation, commissioning, maintenance and repair costs, as well as reducing the vulnerability of the system to copper cable theft. The main potential applications envisaged are for powering, wireless communications, wireless command and control data links, local interlocking and trackside object control, train detection, condition monitoring systems, signals and signalling equipment, route switching (point motors), level crossing equipment (monitoring and actuation), and point heaters. The energy harvesting profiles of the technologies discussed previously are all subject to variation in terms of the energy output from the devices, according to the availability of the energy source they are harvesting energy from. For the power supply (energy harvesting and storage combined) to be reliable, the energy harvester must capture more energy on average than is used, and the storage capacity must be sufficient to accommodate the cycles and variations in energy harvesting and usage, and also include an element of redundancy to ensure reliability.

Modern micro-electronic wireless communication and data links, which could be used for communications and command and control functions, with relatively low data rates have fairly low power consumption (less than 1 W). These could need to be active almost continuously in order to verify the status of equipment with similar latency to conventional wired systems, or power saving procedures, could be implemented when revising railway signalling procedures to be more power efficient. Local signalling control and interlocking, and the command and control of trackside equipment has many features in common with communications and data links in terms of the power consumption, particularly as communications and data links are often an integral part of their function. Current devices used in this application have moderate levels of power usage (tens of Watts) but there is potential for replacing them with modern micro-electronics with power usages an order of magnitude lower.

Train detection systems are used to verify track occupancy so that traffic control systems can safely authorise train movement. Traditional track circuit train detections systems require a moderate amount of power continuously. Axle counters which detect and count each wheel entering or leaving a section of track so that the occupancy of a section of track can be established use less power than track circuits but also need to be responsive and therefore active continuously in current railway signalling practices. Current trackside signals require a low to moderate amount of power continuously (about 10 W), with signals based on LEDs.

The setting of routes on railways is actuated by electrically powered devices (some have a hydraulic power transfer stage) known as point motors. These have a large momentary power requirement, of about 3 kW for 20 s, which occurs intermittently when the route is changed, with an additional low continuous power requirement to confirm the status. Equipment installed at level crossings to prohibit the passage of road traffic ranges from flashing lights, to half or full barriers (with flashing lights as well) and might include a monitoring system. The power consumption of the flashing lights if fairly low and intermittent, the power usage of the barriers is moderate to high and is also intermittent, monitoring systems might be active continuously or only activated when required and would have fairly low power requirements.

In some locations, according to the climate, heaters are installed prevent the moveable rails at junctions from being frozen in place by accumulations of snow and ice. The power usage of these "point heater" devices is very high, in the order of magnitude of kilowatts, is seasonal, and can be active for prolonged periods. Various monitoring systems might be applied to the trackside infrastructure at various locations depending on requirements, these might include noise or landslide detection systems for example. The energy usage of monitoring systems would generally be quite low depending on the type of equipment, and might be very low for devices which are largely passive and only activate a low energy usage process when trains are passing or intermittently, but are otherwise in a dormant state.

3.2 Energy Harvesting Concepts for Trackside Applications

In general, it is envisaged it might be feasible to power applications with very low or low intermittent energy usage might with energy harvesting techniques, which derive their power from the passage of trains, with or without supplemental energy harvesting from a very small solar or wind installation. This could be useful in situations such as tunnels and cuttings, where limited solar radiation or ambient air flow are available. It might be feasible to power applications with low or intermittent moderate energy usage using small installations of wind, solar or combined wind and solar energy harvesting techniques possibly supplemented with additional energy from harvesting techniques which derive their power from the passage of trains. Applications which have large energy requirements would require large solar, wind or solar and wind installations along with a very high capacity energy storage system to make powering them with energy harvesting feasible. Although perhaps technically possible is likely to be impractical and economically unfeasible to power equipment with very large energy requirements, such point heaters, using energy harvesting techniques.

3.3 Development of Trackside Energy Harvesting Solutions

A preliminary analysis and a feasibility study of an trackside energy harvesting solution are necessary development steps in the design of these autonomous systems (Hadas et al. 2018). Mathematical models of the physical principles employed and performance data for commercial energy harvesting systems can be used, in combination with analysis of the energy requirements of the applications, to conduct a feasibility study of useful energy harvesting systems. The waterfall development diagram, shown in Fig. 7, can be used to describe development steps for the design of fully autonomous energy harvesting systems.

3.3.1 Railway Network Parameters

The parameters of the railway environment which affect energy harvesting vary according to the particular geographic characteristics of the railway route and location, railway infrastructure characteristics, and railway traffic characteristics. Variations in particular parameters have different effects on different physical principals of energy harvesting. The geographic conditions of the route mostly affect environmental energy harvesting methods, such as solar and wind, the geographic position of the location affects the solar exposure throughout the year and the pattern of wind conditions. Solar based energy harvesting methods need to be aligned with the sun for optimal energy harvesting although will harvest some energy in other orientations. Also the ground profile and vegetation of the location affect the incident solar and

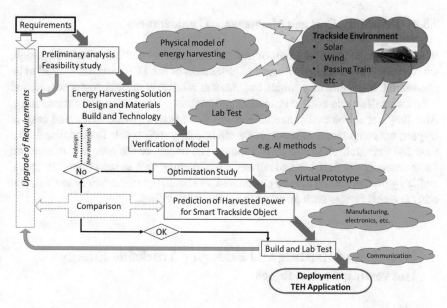

Fig. 7 Waterfall diagram showing the development of energy harvesting system for trackside application

wind energy at the trackside, the local landscape increasing the shading or sheltering the energy harvester compared to an flat open site.

Railway infrastructure characteristics mostly affect energy harvesting from train induced vibration and displacement. Track construction, components and materials; different types of rail, fixing, and rail support (sleepers etc.) affect the dynamic behaviour of the track and therefore the dissipation of the energy imparted by a passing train, which in turn affects the quantity and form (in terms of frequency and amplitude for example) of energy available at a location for the energy harvesters to collect. Also the quality of the maintenance of the alignment affects the interaction between the train and the track and hence the energy input to the track. Other features such as switches and crossings present a disturbance to the support of the wheel, leading to impacts and locally higher energy input into the track. As well as the characteristics of the track, the track support and ground conditions affect the dynamic behaviour of the track.

A factor which affects energy harvesting based on train induced vibration and displacement, and train aerodynamics based wind generators are the railway traffic characteristics, traffic pattern. This includes the length and number of trains, the speed of the trains, the load condition and the vehicle dynamics of the types of vehicles.

3.3.2 Reliability, Cost and Maintenance Considerations

In addition to the physical environment and the practical viability of a TEH design, the reliability, cost and maintenance requirements of the TEH (and track) need to be considered. In general they should be at least as reliable as current power supply, and have lower life cycle costs considering procurement, installation and maintenance. Also they should be easily maintainable at track-side, self-diagnostic and easy to inspect, the components or modules should be easily replaceable to minimise down-time and time staff on site to correct faults. The impact of the energy harvesters on the maintenance of the track and railway infrastructure, such as inspections, tamping, and rail-grinding should also be considered, as should the vulnerability to theft, and other external factors such as extreme winds and flooding.

4 Overview of Existing and Emerging Trackside Energy Harvesting Technologies

4.1 Solar Energy Harvesting Technologies

Various solar panel products are commercially available for general use to supply power to different applications; the range in power output and size varies from very small to panels which have a power output of a couple of hundred Watts and areas of a couple of square meters. Specialist panels could be integrated or applied to structures of various shapes if a particular situation requires it. Small solar panels of around 0.1 m^2 can be applied to equipment or a suitable surface almost anywhere on or around the track and harvest a small amount of power. Larger panels of a couple of square meters could be installed on posts, or similar mountings, next to the track to provide a low to moderate average power output. Larger arrays of solar panels with higher power ratings might require installation on land further from the trackside, possibly outside the normal railway boundary. The performance of commercial solar panels in terms of how much power they generate for each unit of sunlight is well established, e.g., the nominal maximum power rating of crystalline silicone solar panels is approximately 150 W/m^2. The average power output of the solar panels is affected by their orientation, local shading, the geographic location of the installation, seasonal variations, and the weather.

As an alternative to conventional solar arrays the start-up Greenrail recently presented an eco-friendly sleeper, which integrates a photovoltaic (or piezoelectric) module onto a lightweight composite sleeper. Information on the power output or other parameters is not yet available, as the product is still under development; product samples and trial installations are shown in Figs. 8 and 9.

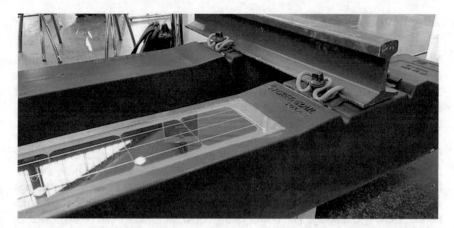

Fig. 8 Solar sleeper (Greenrail Project No. 738373—H2020-SMEINST-2-2016-2017)

Fig. 9 Solar sleeper installation (Greenrail Project No. 738373—H2020-SMEINST-2-2016-2017)

4.2   Wind Turbine Technologies

Various commercially available designs and products for energy harvesting from environmental air flows are suitable for general use supplying power to multiple applications. These conventional wind turbines are most commonly either bladed designs rotating about a horizontal axis, or designs with vanes rotating about a vertical axis, and are available in a variety of sizes with varying power output ratings. Wind turbines of moderate power rating and size, for example power ratings of hundreds of Watts and major dimension of around 1 m, could be installed at most trackside locations, although performance in locations sheltered by buildings or the landscape will be less than optimal. Larger wind turbines with higher power ratings might require installation on land further from the trackside, possibly outside the normal railway boundary. The power ratings of wind turbines can be expressed as either the maximum power output at the optimal wind speed, or more usefully the nominal monthly average power output, although this would vary according to the month, conditions at the installation location in terms of sheltering and exposure, and the profile of the wind conditions at the geographic location throughout the year. Examples of commercially available wind turbines include:

- a three bladed model with a 1.17 m diameter rotor with a swept area of 1.07 m^2, optimum wind speed range is 4.5–22 m/s, and has a 40 kWh/month average output based on an average annual wind speed of 5.8 m/s (Primus Windpower);
- a design with vanes rotating about a vertical axis, which has a rotor of 0.27 m diameter with a height of 0.918 m; it has a momentary power output of 24 W at a wind speed of 8 m/s, and the peak output is 200 W (Leading Edge—Vertical Axis Turbine).

In the context of this paper, devices intended to make use of the aerodynamic effects of passing trains are included in the category of wind turbines. Devices that harvest energy from the aerodynamic effects can either be focused entirely on harvesting energy from the air currents caused by passing trains, or a combination of this and environmental wind. The advantage of the latter approach is that one device can harvest energy from both sources with air currents from trains supplementing he environmental wind, particularly on days where there is little or no environmental wind. By necessity wind turbines which use air currents generated by trains have to be fairly small to fit in the railway environment around the area when trains pass where the air currents are strongest, and consequently of moderate power rating. The disadvantage of these types of wind turbines is that they might not be ideally placed to take advantage of the environmental wind.

An example of an energy harvester focused entirely on harvesting energy from the air currents induced by the aerodynamic effects of passing trains is the T-BOX Wind Power Generator (T-BOX Wind Power Generator). The device, shown in Fig. 10,

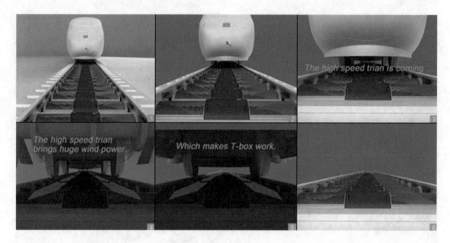

Fig. 10 T-BOX wind power generator (T-BOX Wind Power Generator)

is installed between railway sleepers, and is partially buried underground. As the train passes over the device, the air currents generated from the train aerodynamics spins the turbine inside the T- box to generate electricity. The T- box contains all the mechanical components required for harnessing, storing and supplying converted power. It consists of a durable metallic cylinder with vents, which allow air to flow through and rotate turbine blades housed inside. The Hetronix wind turbine system consists of a 2.5 m long rotor system and a generator which is 35 cm in diameter. The 58 kg wind turbine is rated at 2000 W with a 12.5 m/s airflow.

4.3 Technologies Based on Linear Displacement Electromagnetic Generator Concept

One concept for harvesting energy from the trackside environment is to use the linear displacement induced by the passage of trains and transfer this motion to an energy harvesting component. The displacement could be captured either from contact with the wheel directly or from a connection to the track that is displaced by the train. Some examples of different concepts for capturing energy from train induced displacements are presented further on in this section.

Geared electromagnetic generators prototypes

Most of the concepts and prototypes that have been developed so far are based on devices fitted to the track and operating on the principal of transforming vertical displacement of the track caused by the passage of a train, via racks, gears and clutches, to a rotary motion to drive a rotary electromagnetic generator. The main distinguishing feature between these designs is whether the device is connected to

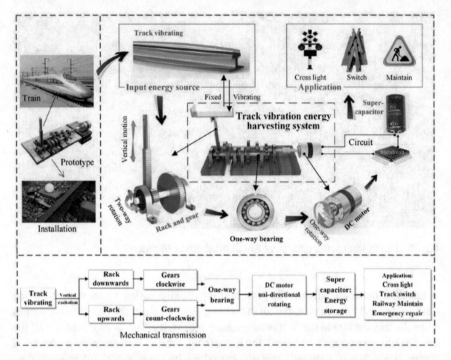

Fig. 11 Design and prototype of the energy harvester with output energy peak voltage of 58 V at 1 Hz with a displacement of 2.5 mm (Zhang et al. 2016)

some form of ground anchor to utilise the relative motion between the track and the ground, or if the device is only connected to the track and utilises the relative motion between different locations on the track.

A team from Southwest Jiaotong University, China, presented a portable high-efficiency electromagnetic energy harvesting system (Zhang et al. 2016), depicted in Fig. 11. It consists of two main parts: mechanical transmission and the electrical regulator. With a displacement of 2.5 mm at 1 Hz, the peak output voltage is 58 V, which is close to being practically useful for supplying trackside applications, such as safety devices and emergency repairs in areas lacking power, indicating that the proposed system has potential as a renewable alternative energy source.

Another prototype of mechanical rectifier based harvester has been developed by a team from Stony Brook (Wang et al. 2012). The results of laboratory testing show that sufficient power can be harvested, as well as the features and benefits of the motion rectifier design, which is shown in Fig. 12.

A novel direct motion-driven harvester has been reported by TTCI (Lin et al. 2018). The harvester shown in Fig. 13 is anchorless and harvests energy without requiring special preparation during installation. Compared to any traditional anchored device, this design is more practical, as it does not require a long interruption to the operation of trains during its installation.

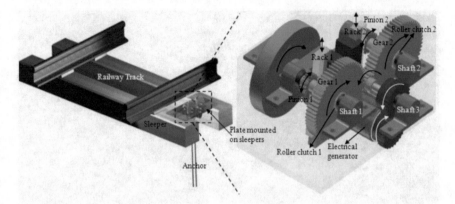

Fig. 12 Electromagnetic energy harvesting from train induced railway track displacement (Wang et al. 2012)

Linear electromagnetic generator concept

Linear generators produce electricity from the relative movement of the components using the electromagnetic effect. They use the same electromagnetic principles as rotary generators. However, linear generators are potentially more suitable for some applications as they do not require complex mechanisms to transfer linear motion into rotary motion to harvest energy. Figure 14 shows a schematic representation of a linear generator.

A novel concept that considers different designs and installation options has been proposed within the EU Horizon 2020 project ETALON by the team at Newcastle University, for integrating linear generators into railway infrastructure. The concept is based on the principle of capturing energy from displacements induced by either the train itself, or the movement of the track under the train and transferring those displacements to a linear generator.

Schematic representations of various potential configurations for linear generator harvesters are shown in Figs. 15 and 16, where the black rectangle in each concept represents the mover of the linear generator, which contains permanent magnets, located within open box representing the stator coil and protective case. In Fig. 15 an actuating arm is connected at one end to the linear generator such that the contact element on the other end is positioned next to the inside of the railhead just below the running surface. When the wheel of a train passes, the flange of the wheel, which extends below the running surface of the rail, makes contact with the contact element and displaces it downwards. This linear (or arc segment) motion is transmitted either directly, as in Fig. 15a, or via a pivot (fixed to the track) to the linear generators (also fixed to the track), with the lever arrangement, as in Fig. 15b, c, amplification of the displacement is possible depending on the length of the levers.

In Fig. 16a, the linear generator and pivot are fixed to the sleeper and the linear generator is connected to the rail through the pivot via a lever to transmit the vertical displacement of the rail relative to the sleeper to the linear generator for energy

Fig. 13 Harvester installed and tested at TTCI test track, with fully loaded freight train running at 64 km/h (40 mph) (Lin et al. 2018)

harvesting. In Fig. 16b the linear generator and pivot are fixed to the ground and the linear generator is connected to the track through the pivot via a lever to transmit the vertical displacement of the track to the linear generator for energy harvesting. In this concept it is expected that there would be a large differential in the lengths of the lever either side of the pivot in order to amplify the small displacement and large forces from the track into a larger displacement at the linear generator.

A prototype has been designed (Fig. 17) for being manufactured and tested within the EU Horizon 2020 project ETALON.

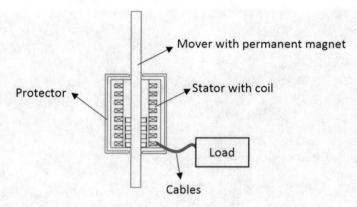

Fig. 14 Schematic representation of a linear electric generator

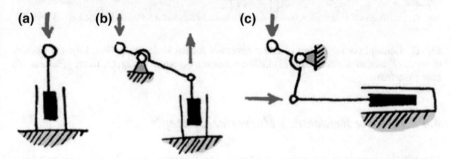

Fig. 15 Schematic representation of linear generator energy harvesters mounted on the track and displaced by the wheel

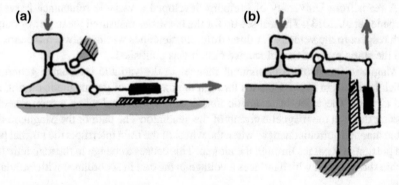

Fig. 16 Schematic representation of linear generator energy harvesters fixed to the ground and displaced movement of the track

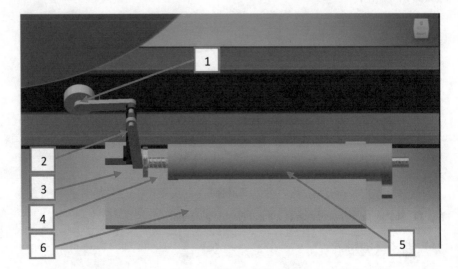

Fig. 17 Concept outline design of linear generator for trackside application. Key: (1) Contact element, (2) actuation mechanism, (3) limit mechanism, (4) return spring, (5) linear generator, (6) base plate/frame

4.4 Variable Reluctance Harvester Concepts

Electrical energy can be generated by an electromagnetic induction, caused by a change in magnetic reluctance induced by a passage of a train wheel, momentarily forming part of the magnetic circuit.

A team from University of Freiburg developed a variable reluctance harvester (Kroener et al. 2013). The test set-up for the harvester measured mean power output with respect to the velocity for three different clearance widths between the moving and the static parts of the reluctance circuit was published.

Magnetic flux from a permanent magnet in the variable reluctance generator, which is shown in Fig. 18, passes through the fixed part of the magnetic circuit, rail and air gap. The value of magnetic flux density is minimal when a train wheel is passing through the magnetic circuit of the generator. The path of the Magnetic flux in the magnetic circuit changes when the wheel of the train interrupts the normal path at a portion of it passes through the air gap. This causes a change in the magnetic flux within the wire coil which induces a voltage in the coil in accordance with Faraday's Law.

A rail concept of the variable reluctance harvester was analysed under the EU Horizon 2020 project ETALON. The simulation results shown in Fig. 19 present the calculated potential energy output of this design. Each passing wheel provides only one positive and one negative peak of voltage in a short time. The magnitude of the voltage peaks is proportional to the train speed and depends on the change of magnetic flux density through magnetic circuit. Due to range of the variation in the

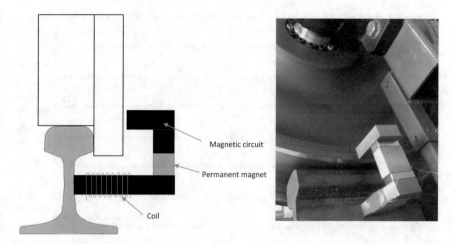

Fig. 18 Variable reluctance rail harvester; model and test concept of ETALON project

lateral position of the wheel relative to the rail, minimal allowed distance between the magnetic circuit and rail to avoid contact between the wheel and the magnetic circuit is 58 mm. Therefore, the change of magnetic flux density due to the presence of a wheel is in range of 100–250 mT, the actual value for any particular wheel pass depending on the actual lateral position of the passing wheel within the range of possible positions.

4.5 Vibration Energy Harvesting Technologies

The applications of vibration energy harvesting technologies are well established in the fields of structural monitoring of civil structures and aircraft. Passing train induce huge mechanical vibrations in the track and trackside environment. Several concepts of trackside energy harvesting technologies were investigated and published. The paper by Cleante et al. (2016) reported on an investigation into how much mechanical energy could potentially be harvested from the vertical vibration of a sleeper induced by trains passing at different speeds. Basic information about the very low outputs of different piezoelectric energy harvesting systems have been reported for a proposed piezo-drum design (Tianchen et al. 2014), a piezoelectric circular membrane array (Wang et al. 2014), piezoelectric vibration cantilever harvester (Gao et al. 2016), and piezoelectric patch-type and stack-type energy harvesters (Wang et al. 2015).

Vibration energy harvesters based on the electromagnetic principal provide more promising source of energy from trackside vibration which could be used for wireless sensor networks. Design of electromagnetic energy harvester by Gao et al. (2018), which converts rail vibrations into electricity, is shown in Fig. 20.

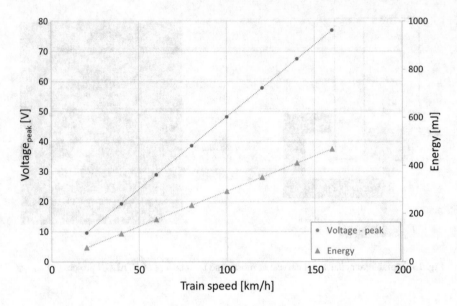

Fig. 19 Variable reluctance rail harvester; simulation results for coil with 2000 turns and an assumed change of magnetic flux density of 100 mT

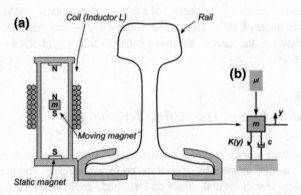

Fig. 20 Electromagnetic energy harvester by magnetic levitation: **a** physical model, **b** representative mechanical schematic (Gao et al. 2018)

The research by Southwest Jiaotong University (Gao et al. 2017a, b) investigated the possibility of establishing a self-powered wireless sensor network by integrating the ZigBee stack protocol together with an energy harvesting power source. Field test of self-sustaining sensor nodes with local energy harvesting is shown in Fig. 21.

Swedish company ReVibe Energy has developed electrodynamic vibration energy harvesters of various sizes. Their pilot project in cooperation with Deutsche Bahn AG includes adaptation of in-house developed inertial energy harvesting units for the trackside environment.

Fig. 21 Field test of self-sustaining sensor nodes with local energy harvesting (Gao et al. 2017a, b); Illustration of hardware prototype of self-sustaining sensor nodes for urban rail transit (Gao et al. 2018)

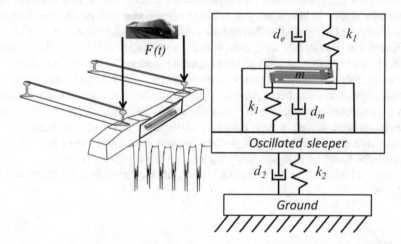

Fig. 22 Principle of pulse excitation by passing train of electromagnetic vibration energy harvester

ReVibe Energy and Southwest Jiaotong University energy harvesting solutions convert vibration of the rail into electricity. The energy harvesting team at Brno University of Technology, Czech Republic, is developing an electromagnetic vibration energy harvester for a sleeper application under the EU Horizon 2020 project ETALON. The principle is shown in Fig. 22 where pulse excitations of the sleeper provide a vibration response of electromagnetic energy harvester.

The cantilever design of electromagnetic harvester is shown in Fig. 23. This system is designed for use with a traditional sleeper design with a simple mounting,

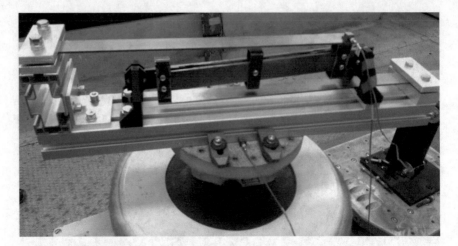

Fig. 23 Shaker test of electromagnetic vibration energy harvester for sleeper

or it could be integrated inside new generation of smart sleepers. This harvester was excited by shaker in laboratory environment using a vibration profile based on real measurement of sleeper vibration and the voltage and power output recorded.

Experimental results of shaker test with the acceleration of sleeper induced by the shaker being representative of a passing regional passenger train are shown in Fig. 24. The harvester power output depends the harvester response to track dynamics, which, in turn, depend on a number of parameters related to the interaction between the track and train dynamics, including the track alignment quality, train speed, weight and suspension type. A freight train passing a harvester of this design mounted on low quality track could provide very high output power peaks (more than 1 W). Whereas, lightweight trains on a high quality high-speed track provide a specific vibration spectrum, which is not suitable for this system, and the response of the harvester will be lower.

5   Conclusion

In conclusion, this paper showed research results that are broadly positive. There are a large number of technologies and systems in development, which have demonstrated the feasibility of powering trackside low-power equipment using energy harvesting.

The identified technologies and potential applications show a wide variety of energy harvesting solutions and energy usage, respectively. Therefore, development of a particular power supply for a specific installation needs to take into account the energy harvesting capacity and characteristics of the energy harvester, along with the energy storage, and the energy requirements of the connected equipment.

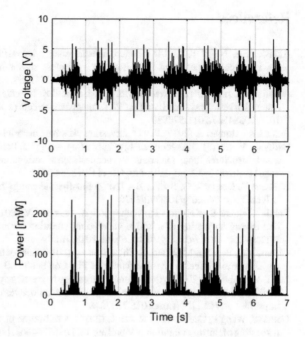

Fig. 24 Experimental results of shaker test of electromagnetic vibration energy harvester. Vibration of sleeper during passing of regional train

Since the energy harvesting capacities of the technologies range from a few microWatts up to a few kiloWatts, there should be a viable solution for most applications, although cost, maintainability, reliability, robustness and redundancy would need to be taken into account. Also the variations in the characteristics of the energy harvesting technologies, such as annual cycles and dependence on passing trains, need to be considered to ensure a reliable power supply is developed, the optimum solution might vary between locations depending on local conditions and requirements. For example, piezoelectric systems are suitable for sensing and monitoring application; electromagnetic resonators seem to be suitable for maintenance free powering of wireless sensor nodes, while variable reluctance harvester could be used for traffic monitoring applications. Displacement linear harvester and geared systems provide interesting source of energy for local electronics. Wind and solar technologies are commonly used renewable sources of energy for wire spectrum of engineering applications.

Hybrid systems harvesting energy from two or more sources and utilising different principles would offer a diversity of energy sources which would increase the robustness of the power source. In general, it was shown that applications with large power requirements would need to at least include a large installation of one or both of solar and wind, combined with large capacity energy storage, in order to meet the requirements. The current technologies which harvest energy from the passage of trains would only be suitable for applications with low power requirements, intermittent moderate power requirements, or as supplemental energy harvesting to other systems.

References

Aktakka EE, Najafi K (2014) A micro inertial energy harvesting platform with self-supplied power management circuit for autonomous wireless sensor nodes. IEEE J Solid-State Circ 49(9):2017–2029. https://doi.org/10.1109/jssc.2014.2331953

Ambrosio R, Jimenez A, Mireles J, Moreno M, Monfil K, Heredia H (2011) Study of piezoelectric energy harvesting system based on PZT. Integr Ferroelectrics 126(1):77–86. https://doi.org/10.1080/10584587.2011.574989

Batra AK, Alomari A (2017) Power harvesting via smart materials. SPIE

Berbyuk V (2013) Vibration energy harvesting using Galfenol-based transducer. In: SPIE smart structures and materials + nondestructive evaluation and health monitoring, pp 86881F–86881F–12. https://doi.org/10.1117/12.2009812

Boisseau S, Despesse G, Seddik BA (2012) Small-scale energy harvesting. Edited by Lallart M. InTech. https://doi.org/10.5772/3078

Cahill P, Hazra B, Karoumi R, Mathewson A, Pakrashi V (2018) Data of piezoelectric vibration energy harvesting of a bridge undergoing vibration testing and train passage. Data Brief 17:261–266. https://doi.org/10.1016/j.dib.2018.01.009

Cleante VG, Brennan MJ, Gatti G, Thompson DJ (2016) Energy harvesting from the vibrations of a passing train: effect of speed variability. J Phy Con Ser 744:012080

Fan T, Yamamoto Y (2015) Vibration-induced energy harvesting system using Terfenol-D. In: 2015 IEEE international conference on mechatronics and automation (ICMA). IEEE, pp 2319–2324. https://doi.org/10.1109/icma.2015.7237848

Gao MY, Wang P, Cao Y, Chen R, Liu C (2016) A rail-borne piezoelectric transducer for energy harvesting of railway vibration. J VibroEng 18(7):4647–4663. https://doi.org/10.21595/jve.2016.16938

Gao M, Lu J, Wang Y, Wang P, Wang L (2017a) Smart monitoring of underground railway by local energy generation. Underground Space 2(4):210–219. https://doi.org/10.1016/j.undsp.2017.10.002

Gao M, Wang P, Wang Y, Yao L (2017) Self-powered ZigBee wireless sensor nodes for railway condition monitoring. IEEE Trans. Intell. Transport. Syst. 1–10. https://doi.org/10.1109/tits.2017.2709346

Gao M, Li Y, Lu J, Wang Y, Wang P, Wang L (2018) Condition monitoring of urban rail transit by local energy harvesting. Int J Distrib Sens Netw 14(11):155014771881446. https://doi.org/10.1177/1550147718814469

Hadas Z, Vetiska V, Huzlik R, Singule V (2014) Model-based design and test of vibration energy harvester for aircraft application. Microsyst Technol 20(4–5):831–843. https://doi.org/10.1007/s00542-013-2062-y

Hadas Z, Janak L, Smilek J (2018) Virtual prototypes of energy harvesting systems for industrial applications. Mech Syst Signal Process 110. https://doi.org/10.1016/j.ymssp.2018.03.036

Jiang T, Chen X, Yang K, Han C, Tang W, Wang ZL, (2016) Theoretical study on rotary-sliding disk triboelectric nanogenerators in contact and non-contact modes. Nano Res 9(4):1057–1070

Kaleta J, Kot K, Mech R, Wiewiorski P (2014) The use of magnetostrictive cores for the vibrations generation and energy harvesting from vibration, in the selected frequencies of work. Key Eng Mater 598:75–80. https://doi.org/10.4028/www.scientific.net/KEM.598.75

Kroener M, Ravindran SKT, Woias P (2013) Variable reluctance harvester for applications in railroad monitoring. J Phys Conf Ser 476:012091. https://doi.org/10.1088/1742-6596/476/1/012091

Lin T, Pan Y, Chen S, Zuo L (2018) Modeling and field testing of an electromagnetic energy harvester for rail tracks with anchorless mounting. Appl Energy 213:219–226. https://doi.org/10.1016/j.apenergy.2018.01.032

Mateu L, Moll F (2005) Review of energy harvesting techniques and applications for microelectronics. In: Lopez JF, Fernandez FV, Lopez-Villegas JM, de la Rosa JM (eds) VLSI circuits and systems II, Pts 1 and 2. Proceedings of the Society of Photo-Optical Instrumentation Engineers (Spie), pp. 359–373. https://doi.org/10.1117/12.613046

Tianchen Y, Jian Y, Ruigang S, Xiaowei L (2014) Vibration energy harvesting system for railroad safety based on running vehicles. Smart Mater Struct 23(12):125046. https://doi.org/10.1088/0964-1726/23/12/125046

Wang W, Huang R-J, Huang C-J, Li L-F (2014) Energy harvester array using piezoelectric circular diaphragm for rail vibration. Acta Mech Sin 30(6):884–888. https://doi.org/10.1007/s10409-014-0115-9

Wang JJ, Penamalli GP, Zuo L (2012) Electromagnetic energy harvesting from train induced railway track vibrations. Proceedings 2012 8th IEEE/ASME international conference on mechatronic and embedded systems and applications. MESA 11787:29–34. https://doi.org/10.1109/MESA.2012.6275532

Wang J, Shi Z, Xiang H, Song G (2015) Modeling on energy harvesting from a railway system using piezoelectric transducers. Smart Mater Struct 24(10):105017. https://doi.org/10.1088/0964-1726/24/10/105017

Yoon Y-J, Park W-T, Li KHH, Ng YQ, Song Y (2013) A study of piezoelectric harvesters for low-level vibrations in wireless sensor networks. Int J Precision Eng Manuf 14(7):1257–1262. https://doi.org/10.1007/s12541-013-0171-2

Zhang L, Jin L, Zhang B, Deng W, Pan H, Tang J, Zhu M, Yang W (2015) Multifunctional triboelectric nanogenerator based on porous micro-nickel foam to harvest mechanical energy. Nano Energy 16:516–523. https://doi.org/10.1016/j.nanoen.2015.06.012

Zhang X, Zhang Z, Pan H, Salman W, Yuan Y, Liu Y (2016) A portable high-efficiency electromagnetic energy harvesting system using supercapacitors for renewable energy applications in railroads. Energy Convers Manag 118:287–294. https://doi.org/10.1016/j.enconman.2016.04.012

Zi Y, Niu S, Wang J, Wen Z, Tang W, Wang ZL (2015) Standards and figure-of-merits for quantifying the performance of triboelectric nanogenerators. Nat Commun 6(1)

High-Speed Overnight Trains—Potential Opportunities and Customer Requirements

Bernhard Rüger and Peter Matausch

Abstract The European high-speed network already offers an alternative to intra-European air traffic for short distances (500–1000 km). High-speed traffic has so far been limited to daily connections. By making targeted use of the overnight jump, the train could also be an alternative to air travel for distances of up to 2000 km. This paper shows the potential opportunities for the use of high-speed overnight trains in Europe and defines the requirements for services and vehicle equipment from the point of view of rail passengers.

Keywords High speed · Rail · Passenger demand · Overnight trains

1 Introduction

Compared to high-speed trains running during the day, conventional night trains often use obsolete rolling stock. Furthermore, in many cases the speeds are below those of day-train traffic, which results in long travel times for relatively short distances and often means that the destination is not reached even after the night's rest. Air traffic presents strong competition to rail transport because of relatively low ticket prices. This has led to reducing night traffic through decreasing demand. At the same time, high-speed transport is being developed in Europe.

The new construction of lines for correspondingly higher speeds and the use of modern high-speed trains result in correspondingly shorter travel times between European metropolises. In many cases, competition for conventional night-train traffic has also contributed to a reduction in night-train traffic.

This paper addresses the question of whether it is possible for railway infrastructure to set up high-speed night train services in Europe. The question arises as to whether it will be possible in the future to cover long distances in a reasonable time

B. Rüger (✉)
Research Centre for Railway Engineering, Vienna University of Technology, Vienna, Austria
e-mail: bernhard.rueger@tuwien.ac.at; bernhard.rueger@fhstp.ac.at

B. Rüger · P. Matausch
St. Pölten University of Applied Sciences, Sankt Pölten, Austria

© Springer Nature Switzerland AG 2020
M. Marinov and J. Piip (eds.), *Sustainable Rail Transport*, Lecture Notes in Mobility,
https://doi.org/10.1007/978-3-030-19519-9_9

during the night hours with the infrastructure already in place, and the infrastructure planned or under construction. Furthermore, the paper examines on which of these possible potential connections there is already a high passenger volume in air traffic and thus the high-speed night train can appear as a possible direct competition to it.

2 Principles and Requirements for Night Train Services

Before the potential opportunities of high-speed night traffic are discussed, some principles and requirements are presented. In the night train, sleeping compartments, couchettes and sleeperettes are used for travel, in addition to seating coaches, as customers spend most of their time asleep. There are different requirements both for the coach material (comfort) and the personnel as well as for the attractiveness of the arrival and departure times. In addition, as there are many differences in the railway infrastructure in Europe that have evolved over time, the problem of interoperability arises due to the long distances covered by night trains. It is also important to consider the possibilities offered by railway-coach connections and whether these should be used for high-speed night trains in the future (Hödl 2006).

2.1 Departure and Arrival Times

Attractive departure and arrival times are crucial for the use of night trains. In the course of a survey undertaken for a research project at Institute of Transport Economics and Logistics of the Vienna University of Economics and Business Administration, travellers in night trains were asked, among other things, about the importance of various reasons for their decision to travel by night train. Eighty percent of the respondents indicated that departure and arrival times are 'very important' or 'important'. The only considerations more important for the respondents were the use of the night as travel time as well as safety while travelling and cleanliness (Hödl 2006).

In order to analyse the importance of departure and arrival times, in a research project referring to improve attractiveness of over-night trains various reasonableness categories are used for business travellers. The categorisation used includes: travel duration, the time without departures, transfers and arrivals (the time the train travels without stopping) and the maximum number of transfers. Figure 1 shows the various reasonableness categories (Binder 2011).

Night-train connections are described in category Reasonableness 1. This category identifies 20:00 in the evening as the preferred departure time. The arrival time in the morning should be between 06:00 and 08:00. A continuous travel time of at least six hours with no transfers are also important for the acceptance of a night train. It is not possible for operators to position transfers in night trains on the market and therefore, transfer-free travel has high priority. In order to be able to survive in competition with

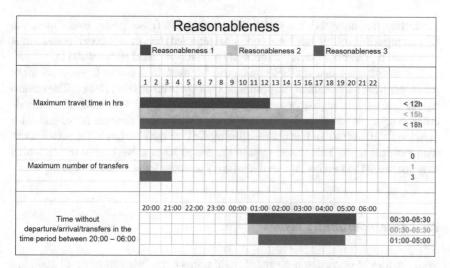

Fig. 1 Three-part classification for the reasonableness of rail connections within Europe

air travel, the travel time itself is not so decisive for the attractiveness of a connection as long as it lasts a maximum of 12–14 h.

Reasonableness 2 refers to the number of transfers a passenger will accept. A one-time transfer one-time transfer and is still assessed positively in principle, as it is quite reasonable to travel with another train after the night connection as long as the total travel time is not too long. The maximum one-time transfer is described as the upper limit of reasonableness for connections between European capitals. In this case, the connections must function optimally. Another option to increase the reasonableness would be to extend the time without departures, transfers and arrivals.

Travel that falls into category 3 is no longer relevant for the night-train market because of poor attractiveness referring to a very long travel duration and the number of transfers. In most cases customers choose the aircraft for such connections (Binder 2011).

2.2 Possibilities by Means of Through-Coach Connections

By using through coaches or portioned trains, parts of the train may have varying destinations or departure points. Through coach means that one or more coaches with different destinations are attached to different trains. Portioned trains means that after one portion of the route, the train is usually divided into two parts, and then the individual parts each proceed to a different destination station. Conversely, this concept is also possible if two trains from different source stations are brought together to reach the destination station. The advantage of the system of portioned trains and through coaches is that it is also possible to provide areas with a direct connection that are not directly on a night-train route (Hödl 2006). In addition, the cost savings (less rolling stock and personnel required) on through-coach connections also make it possible to serve connections with low demand (Binder 2011).

If it is assumed that multiple units will be used in a future high-speed night passenger train service, then the portioned concept is an obvious choice. The advantage is that modern high-speed multiple units are equipped with automatic central-buffer couplings, which makes the coupling process much simpler, faster and safer (Wikipedia 2014/03/13). It is not possible to use through coaches in the actual sense with multiple units. This would require the use of locomotive-hauled trains, which may not reach the maximum speed necessary to cover longer distances at night.

2.3   Interoperability Issues

With the aim of operating high-speed night trains over distances of up to 2000 km within Europe, the question of the necessary interoperability inevitably arises. Since the railway network in Europe consists of many historically grown railway systems, the different technical standards make cross-border rail traffic more difficult. Differences between European Union Member States in vehicles, technology, signalling, safety regulations, braking systems, types of traction current and speed limits mean that trains in international traffic have to stop at "borders". This limits the competitiveness of the railways and slows down traffic, a feature that does not appeal to customers who would potentially use night trains.

Four directives have been published by the European Union with the aim of reducing existing differences, of which the two oldest have already lost their validity and have been replaced by a new directive:

- Directive 96/48/EC on the interoperability of the trans-European high-speed rail system (expires 19.07.2010)
- Directive 2001/16/EC on the interoperability of the trans-European conventional rail system (expired 19.07.2010)
- Directive 2004/50/EC amending the above mentioned directives.
- Directive 2008/57/EC on the interoperability of the rail system within the EU Community.

Directive 2004/50/E updates the existing legislation on high-speed rail systems and extends Directive 2001/16/EC to the whole European-rail network. The first step in the area of interoperability was taken with the adoption of Directive 96/48/EC. In order to achieve the proposed objectives, drafts of technical specifications for interoperability (TSIs) have been drawn up by the European Association for Railway Interoperability, which represents the interests of infrastructure managers, railway companies and industry. The TSIs serve as technical solutions to meet the essential requirements of interoperability and are also intended to contribute to the operational capability of the railway system. Directive 2001/16/EC introduced collaborative procedures for the development and adoption of TSIs for both conventional and high-speed rail. (Europe Legislation 2014/03/13)

An example of a TSI is the European Rail Traffic Management System (ERTMS), which when finally deployed will bring benefits in terms of safety, performance, punctuality and reliability (Rambausek 2009).

The latest Directive 2008/57/EC now serves as the legal basis for the TSIs and was created as a replacement for Directives 96/48/EC and 2001/16/EC in order to harmonise the rules for high-speed and conventional rail in one directive (BAV 2014/05/15). The objective of Directive 2008/57/EC is to accelerate the integration of the railway network of the European Union through technical harmonisation (Europe Legislation 2014/05/14).

With regard to high-speed night trains, it is to be hoped that the construction and planning of international connections of high-speed lines and the resulting expansion of international traffic will enable European regulations to take effect and eliminate the lack of interoperability (Rambausek 2009).

2.4 Requirements for the Coach Material

In contrast to day trains, night trains have three to four different classes, which can be chosen according to the needs and financial means of the passengers. These include the classic seating coach, the sleeping coach and couchette coach as well as sleeperettes (sleeping armchairs) on some connections.

Seating coach:

These are used as large-capacity coaches or as compartment coaches for six persons, in some countries also for eight persons (e.g. Spain) and are often air-conditioned. There is usually no surcharge for the seating coach, but a reservation is often required (Hödl 2006).

Sleeperettes:

Sleeping armchairs are usually operated as large-capacity coaches and meet the rest needs of travellers more than classic seating coaches, also because the light is often almost completely extinguished at night. However, the sleeperettes were not well received by passengers because they prefer the comfort of a flat bed. Therefore they were replaced with couchettes by many railway companies (Hödl 2006).

Couchette coach:

These are used as compartment coaches for four or six persons with simple berths. The compartment can still be used in the evening hours as a classic seating compartment and only before the night's rest will the berths be folded out either by the passengers or by the passenger attendant. The berths are equipped with sheets, blankets and pillows. The sanitary facilities usually consist of a central washroom and toilets, but there are no showers. There is a care service in the form of a wake-up service and in some instances a small breakfast (Hödl 2006).

Sleeping coaches:

These are used as compartment coaches with one to three folding beds and offer a hotel-like service. Each compartment has a washing facility, and some have their own shower and toilet. There is care service in the form of a wake-up service, breakfast in bed up to small meals in the compartments (Hödl 2006).

The CRH1E electric railcar in China, which is based on the platform of the Bombardier ZEFIRO 250, shows what the future development towards high-speed night trains could look like. This is described by Bombardier as the world's fastest sleeper train and is currently in service on the Beijing-Shanghai route. The 16-part train reaches a top speed of 250 km/h and consists of a luxury sleeping coach, 12 standard sleeping coaches, a dining coach and two second-class seating coaches (Wikipedia 2014/03/15).

2.5 General Requirements for Passengers on Night Trains

The best possible use of travel time is essential for the attractiveness of night trains in comparison to other means of transport such as air travel. In general, the big advantage of using night trains is the so-called "night jump". This means that the travelling of, sometimes, large distances happens while sleeping. When using an airplane in order to be able to attend a meeting at the destination in the morning, travel must begin the evening before, and then the night must be spent in a hotel at the destination. Or alternatively, the first flight must be taken very early in the morning, which also entails major restrictions on comfort. However, in order to be able to use the travel time efficiently, a variety of requirements must be considered. While for travellers who opt for seating coaches, sleeperettes or couchettes, the main focus is on the fare. While for travellers who use sleeping coaches, much higher demands are on the usability of the travel time and thus the equipment of the vehicles.

In general, the smoothest possible running and a reduction of noise emissions and noise levels are important for all coach types. It is also important for all travellers that their luggage can be sufficiently accommodated. On average, it can be assumed that one piece of luggage per person is taken on the night train (Rüger 2004). This corresponds approximately to the size of normal checked luggage in air travel. As far as accommodation is concerned, it should be taken into account that there are actually sufficient luggage racks per person and therefore per piece of luggage. Please note that only about 20% of travellers are prepared to, and a maximum of 50% of travellers are able to, put their luggage in an overhead storage! (Plank 2007).

For travellers in sleeping coaches it is important to be able to use the travel time before sleeping in the evening as well as after getting up in the morning in a meaningful and efficient way. The same requirements then apply for night travel as in day traffic. For example, for business travellers it is important to have a seat and a table in the compartment in order to be able to work. It is important for all travellers to have a power supply for the use of technical devices. Since more than one device is usually used, at least two sockets per person should also be available in the compartment. A well-functioning internet is also expected along the entire route; this must work across borders especially for trains in international service (FLEXICOACH 2012). Not all travellers expect their own shower and toilet in the compartment, but these days they are part of an appropriate standard. At least half of the compartments should be equipped with a shower and WC.

3 Potential Opportunities in the European Infrastructure Network 2025+

In order to determine the potential for a high-speed night train service, it is necessary to present an overview of the existing infrastructure and the infrastructure under construction or planning, which enables the corresponding traffic. The main bases for assessment are the corridors of the Trans-European Transport Network (TEN-T) and the European Transport White Paper. On the basis of the future intra-European infrastructure network, possible travel times and exemplary connections for the high-speed night train will be investigated.

The 2011 Transport White Paper of the European Commission sets out the vision for European transport in 2050 and a strategy for its implementation. One of the objectives of this vision is to triple the length of the existing European high-speed network by 2030. According to the White Paper on Transport, the creation of a single European transport area is crucial for strategic implementation. The remaining barriers between modes of transport and national systems will be removed in order to facilitate the integration process and encourage the emergence of multinational and multimodal operators. There are major challenges in the internal market for rail transport services, where technical, administrative and legal barriers need to be removed to facilitate entry into national rail markets (White Paper 2011).

For the implementation of a single-transport area, there are corridors defined by the European Union, which together form the TEN-T networks (Trans-European Network Transport). The TEN-T Community guidelines for the development of the trans-European transport network serve as a framework for the development of internationally important transport infrastructure within the EU by 2020. There are currently 30 priority projects with completion planned by 2020, of which 18 are railway projects. The objective of the railway corridors is, inter alia, the interconnection of 15,000 km of railway lines designed for high-speed, the reduction of bottlenecks with the realisation of 35 cross-border projects and the connection of 38 major airports by rail to the conurbations (Europe 2014/04/10).

The travel times possible in the European infrastructure network in 2025 depend on the extent to which the national and international railway projects are realised on the basis of the TEN-T. In order to get an overview of the possible future connections, the future travel times from the three European cities Paris, Frankfurt am Main and Vienna to other European cities were determined in the course of this work. In addition to the central location of these three cities, Paris and Frankfurt am Main are the second and fourth largest airports in terms of passenger volume in the European Union (Wikipedia 2018/10/17).

Due to the central location of the cities, the accessibility in the different regions of Europe can be well represented and through the passenger data of the airports (e.g. passengers by destination, flight offers by destination) potential opportunities for possible future connections are represented. The possible travel times are shown starting from the respective city with arrows in different colours. Each colour means a different travel time. Travel times from approx. eight hours to approx. fourteen hours are shown. If new lines or upgraded lines are already in operation or no new lines or upgrades are planned until 2025 and therefore the current travel times are decisive, the travel times are determined by means of an online query of timetables of different European railway companies (OeBB 2018). If new or expansion projects are known on the routes, the travel times are determined via the various information pages of infrastructure projects, infrastructure managers and railway companies, which indicate the travel times planned for the future on their routes. The travel times are not calculated on the basis of the future maximum speed specified for the route in question but are purely information from project operators and railway companies on the future travel time.

Between the Western European countries of France, Spain, Italy, Germany, Great Britain and the Benelux countries, there are potential opportunities with travel times under 12 h. This is due to the progressive development of domestic high-speed networks in these countries and the development of international high-speed links between these countries (Binder 2011). In 2009, 5500 km of high-speed lines were in operation in France, Italy, Spain and Germany and a further 8411 km were under construction or planning (Rambausek 2009).

However, for the travel times given in Figs. 2, 3 and 4, some decisive projects still need to be realised, including the expansion of international connections between the individual high-speed systems. There are already individual examples of international connections today. Since the opening of the last section for passenger transport between Barcelona and Figueres in 2013, Spain and France have had a cross-border high-speed line (UIC 2014). The travel time between Paris and Barcelona is now only 6 h and 25 min. Another example is the Eurotunnel between France and Great Britain, which provides a travel time of 2 h and 17 min between Paris and London. Further international high-speed connections are already under construction or in the planning stage. These include the new line between Lyon and Turin, including the Monte-Cenis Base Tunnel, which will enable a travel time of two hours instead of the previous four hours between these two cities (Railway Technology 2014). Further examples are the Brenner Base Tunnel between Italy and Austria including access routes, which should enable a travel time of three hours between Munich and

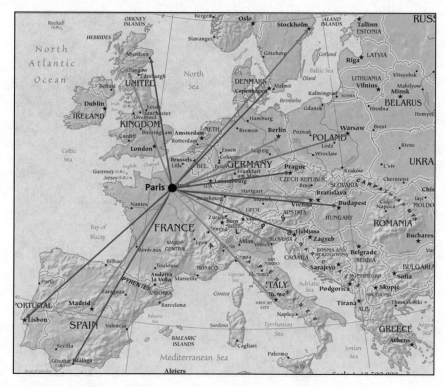

Fig. 2 Representation of possible travel times 2025 from Paris; green ≈ 8 h orange ≈ 10 h purple ≈ 12 h red ≈ 14 h

Verona (BBT 2014) and the Fehmarnbelt Tunnel, which represents a new connection between Germany and Denmark under the Baltic Sea and should reduce the travel time from Hamburg to Copenhagen to three hours (Fermers 2014).

Work on another project, the Madrid-Lisbon high-speed line, has been suspended for the time being due to the financial crisis, and completion by 2025 is questionable. (Caboruivo 2014) Another cross-border project is the modernisation of the Budapest-Belgrade link. The travel time on the 400 km route is to be reduced from the current eight hours to just four hours (Serbian railways 2014). In addition to these cross-border projects, there are several other decisive national projects that will lead to future travel time reductions. France and Spain, in particular, are planning and building high-speed lines, but projects are also underway or planned in Germany, Italy, Poland and Sweden (Wikipedia 2018/10/17).

Obviously, there is a strong west-east disparity in future travel times. According to the forecast maps for the high-speed network in Europe in 2025, the high-speed lines will be located in Western Europe, as is currently the case, and therefore longer travel times are to be expected in Eastern European countries in the future as well (Rambausek 2009).

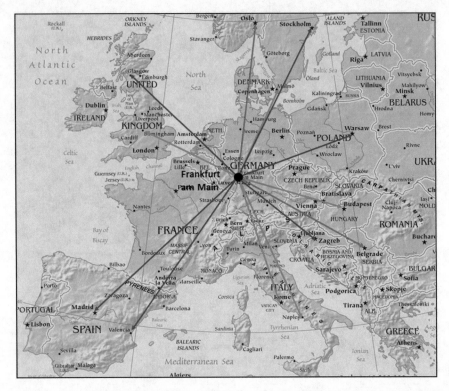

Fig. 3 Representation of travel times 2025 from Frankfurt/Main; green ≈ 8 h orange ≈ 10 h purple ≈ 12 h red ≈ 14 h

Figures 5 and 6 illustrate which distances [air-line distance according to the online distance calculator (air-line 2018)] can be covered at which travel times from Paris and Frankfurt am Main in 2025. The following relations serve as an example of the great difference in travel times between Western and Eastern Europe: the future possible travel time from Frankfurt am Main to Madrid (distance as the crow flies 1446 km) is approx. 12 h, while the future possible travel time from Frankfurt am Main to Bucharest (distance as the crow flies 1455 km) is approx. 23 h.

When using high-speed lines at night, any operational restrictions must also be taken into account. For example, there are high-speed lines designed for mixed operation, whereby the line is used by both passenger and freight trains. Mixed operation often results in reduced line performance and the timetable design is very complicated. Therefore, on mixed routes freight transport is often limited to the night, which inevitably results in lower maximum speeds even for night trains. In addition to use by freight trains, there may also be restrictions due to maintenance work and inspections at night (UIC 2013).

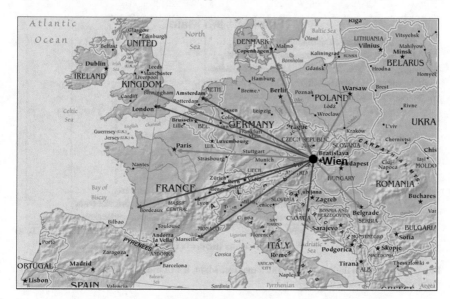

Fig. 4 Representation of travel times 2025 from Vienna; green ≈ 8 h orange ≈ 10 h purple ≈ 12 h red ≈ 14 h

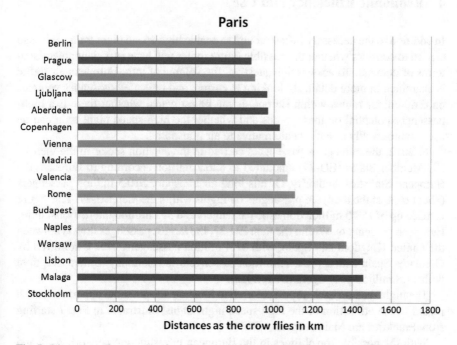

Fig. 5 Distances as the crow flies from Paris, travel times: green ≈ 8 h orange ≈ 10 h purple ≈ 12 h red ≈ 14 h

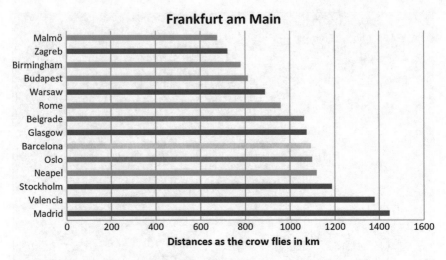

Fig. 6 Distances as the crow flies from Frankfurt/Main, travel times: green ≈ 8 h orange ≈ 10 h purple ≈ 12 h red ≈ 14 h

4 Economic Efficiency and Use

In addition to the necessary infrastructure to enable high-speed night trains, it is also crucial to consider whether the possible future routes will have sufficient potential in terms of demand. To answer this question, the volume of intra-European air traffic is examined in more detail. By looking at current and projected passenger volumes on different air routes within Europe, it can be estimated whether there is a high-passenger potential on these routes and whether the high-speed night train can be used economically as a direct alternative to air transport.

In 2012, the number of passengers carried in the aviation sector in the then 27 EU Member States (EU-27) amounted to 826.7 million according to Eurostat, the European Statistics Authority. Of this total air transport, 510.5 million passengers (about 62% of the total) are passengers on flights within the then EU-27. This figure is made up of 159.5 million domestic passengers and 351 million international intra-European passengers. Within the then EU-27, the largest passenger flow is between the United Kingdom and Spain, with 31.4 million passengers. This is followed by Germany–Spain with approx. 22 million passengers and Germany–United Kingdom with 11.8 million passengers (DLR 2012).

The flight data for each airport are used to illustrate the potential opportunity with regard to the utilisation of the high-speed night trains, illustrated in Fig. 7 starting from Frankfurt am Main Airport.

With the possible travel times in the European infrastructure network 2025 and the passenger data from European air traffic, various exemplary connections can be represented. An example of such a connection can be found in the UIC study on

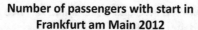

Number of passengers with start in Frankfurt am Main 2012

Denmark	411598
Sweden	453920
France Rest	462010
Greece/Cyprus	467455
Portugal	496974
Poland	502352
France Paris	532610
Switzerland	660261
British Isles Rest	718678
British Isles London	905921
Austria	953899
Spain Continent	957985
Italy/Malta	1580340

■ Number of passengers per year

Fig. 7 Passengers on board for departures from Frankfurt Airport 2012 for flights within Europe (scheduled and charter flights) (Masuch 2014)

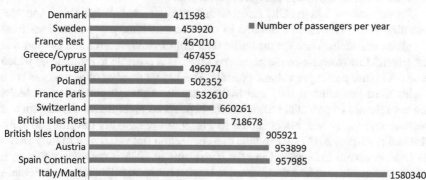

Boarding area	Night time	De-boarding area
Night Train stops at several origins	Nonstop run to cover distance	Train stops at several destinations
▌ Boarding time of high quality Night Trains usually ends at midnight ▌ Boarding after midnight is not attractive ▌ Travel time 0 – 3 hours ▌ Travel distance 0 – 400 km	▌ Defined boarding and de-boarding areas lead to a quiet night time for the passengers ▌ Travel time 6 – 12 hours ▌ Travel distance 1,100 – 2,200 km	▌ De-boarding starts not before 6 a.m. due to attractive time level ▌ Travel time 0 – 3 hours ▌ Travel distance 0 – 400 km
Expample: ▌ London ▌ Lille ▌ Paris	Night time between Paris and Barcelona	▌ Barcelona ▌ Zaragoza ▌ Madrid

Fig. 8 Example of a night train connection from Great Britain to Spain (UIC 2013)

high-speed night trains from 2013 in which a connection from Great Britain to Spain is presented as an example (see Fig. 8).

The "boarding area" is the period during which the train stops at one or more stations so that passengers can board the train. The travel time of this area is specified in the study between zero and a maximum of three hours, also depending on the number of stops. The distance covered is between 0 and 400 km. The next area is the "night time", where in order to give the passengers a night's rest, the train travels through without stopping. It will take six to twelve hours to cover 1100–2200 km. The last area is called the "de-boarding area", where the train stops at only one or more selected stations, as in the first area, to allow passengers to disembark. The travel time should also be between zero and a maximum of three hours and the distance travelled between 0 and 400 km (UIC 2013).

The scheme used in the UIC study, with "boarding area", "night time" and "de-boarding area", can also be applied to other connections, e.g. connections from Frankfurt am Main. With the air traffic data from Frankfurt am Main Airport and the general European air-traffic passenger flows, it is possible to estimate on which routes a higher passenger volume prevails. The largest number of passengers is on flights from Frankfurt to Italy and Malta (Fig. 9). Although passengers to Malta are not relevant as potential customers for high-speed night trains, the majority of passengers to Italy and Malta shown in Fig. 9 are passengers travelling to Italy. (Frankfurt Airport 2014) An example of a connection from Germany to Italy would be from Frankfurt to Naples. Here the travel time in 2025 is only about 8 h, so it seems reasonable to extend the distance to Brussels or Amsterdam in order to achieve a suitable travel time in north-south relation. A train could start at eight o'clock in the evening in Amsterdam and after a journey of 2 h 44 min would be in Cologne and another hour in Frankfurt. After a further approx. six and a half hours the train would then be at approx. seven o'clock in the morning in Rome and after a further 1 h 18 min travel time in Naples. Florence or Bologna would be a possible stopover in the morning.

Although most passengers fly from Frankfurt to Italy and Malta, the strongest passenger flows at the federal level are between Germany and Spain. In 2012, approximately 957,000 passengers took off from Frankfurt for Spain, excluding the Balearic Islands and the Canary Islands. An exemplary connection is offered from Frankfurt via Paris to Barcelona and Madrid. In the future, the journey time between Frankfurt and Madrid would be approximately 12 h (see Fig. 10).

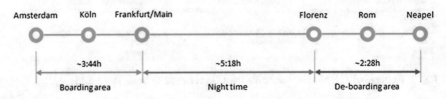

Fig. 9 Exemplary Amsterdam-Naples connection

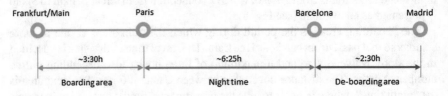

Fig. 10 Exemplary connection Frankfurt/Main-Madrid

5 Conclusion

Due to the already existing high-speed lines in many countries of the European Union, the potential for high-speed night trains already exists today. However, there are strong differences between Western and Eastern Europe. For example, in Western European countries such as Spain, France, Italy, Germany, the Netherlands and Belgium dense high-speed networks and national programmes for further expansion have been in place for years, resulting in a corresponding reduction in travel times. In contrast, in the EU Member States such as Romania, Bulgaria, Croatia, the Czech Republic, Hungary, Greece as well as the Balkan states, travel times in terms of distance travelled are far longer than in Western Europe. Even though the further expansion of the high-speed network is the declared goal of the European Union, it is currently largely restricted to Western Europe. From today's point of view, at least until 2025, no significant reduction in travel time can be expected on Eastern European connections and therefore, the potential for high-speed night trains in that area is rather low. An exception is Poland, where there are already plans for the construction of a high-speed network. Although a large network of high-speed lines already exists in Western Europe, it is mainly cross-border high-speed links between these networks that are still lacking. There are already connections between France and Spain and between France, the Netherlands, Belgium and England. However, other important high-speed links are under construction or in the planning stage. These include: the Brenner Base Tunnel with access routes, the route from Lyon to Turin with the Monte-Cenis Base Tunnel, the Fehmarnbelt crossing and the connection between Madrid and Lisbon. After the completion of these and other crucial projects, which largely correspond to the corridors of the TEN-T, there is potential for many intra-European relations. Connections up to 1400 km could be covered within 12 h (category 1 of reasonableness), e.g. the connection Frankfurt-Madrid, Amsterdam-Naples or Paris-Lisbon. The London-Madrid relation shows that some of these connections are already possible today. A continuous network of high-speed lines would currently allow a travel time of less than 12 h. The necessary interoperability must be ensured for cross-border connections. It is particularly important here that the interoperability directives laid down by the EU are implemented accordingly and that there are no national differences in the high-speed network. Differences between national networks may lead to an extension of travel time through possible necessary border stops. Possible operational restrictions on the high-speed lines due to maintenance or mixed operation at night must in any case be observed but should not lead to any significant restrictions if planned accordingly.

In air transport, the largest flows within the EU Member States are between the United Kingdom, Spain, Germany, Italy, France, the Netherlands and Belgium. Based on flight data from Vienna and Frankfurt airports, it can be seen that there are seasonal differences in the offers for different routes. On the Frankfurt-Madrid route, for example, approx. 1400 seats per day are offered in July and approx. 650 per day in January. If the annual passenger traffic is calculated on one day, an average of approx. 2600 passengers per day fly from Frankfurt alone to the Spanish mainland.

Germany-wide it is approx. 11,000 per day. Even on lower-volume routes such as between the United Kingdom and Portugal, an average of 7000 passengers fly daily in one direction. On the routes where there is a potential due to the high-passenger flows, it is desirable that part of the passengers use the high-speed night train instead of the aircraft.

Potential customers are, for example, business travellers who could save themselves a flight and a night in a hotel by using the train. Another important group are people who are afraid of flying, especially those who only board a plane when there is no alternative. Ultimately, it is also a question of environmental awareness. Many customers could be motivated to switch to the high-speed night train by arguing that CO_2 savings can be achieved by avoiding flights. An essential advantage of the train compared to the airplane is shown by the exemplary connections. With the high-speed night train, it is possible to plan as many stops as desired within the time frame of the "boarding area". Thus, the passengers can be "collected" from the area. This possibility almost does not exist when flying. The same applies to the time frame of the "de-boarding area" where the passengers are distributed back into the area. The major advantage of an overnight train compared to air travel is that destinations away from large airports can be reached directly and not, as in many cases with the use of airplanes, the means of transport has to be changed again in order to reach the start or target destination. By train, you can reduce the number of journeys to and from the airport.

In summary, it can be said that there is potential for intra-European high-speed night trains from the infrastructure side, and that this potential will increase as the network expands by 2025 and beyond. However, most of this potential is currently limited to Western European countries.

Even though concrete opportunities, with regard to possible passengers, still need to be examined separately, the current passenger figures for various destinations within Europe in any case suggest that a corresponding supply of high-speed night trains will be accepted and that there will be sufficient demand.

References

Binder (2011) Klimafreundliche Arbeitsmobilität in Europa/ Charakterisierung und Darstellung von Nachtzugverbindungen innerhalb Europas als Entscheidungshilfe für klimafreundliche Arbeitsmobilität (engl.: Climate-friendly labour mobility in Europe/Characterisation and presentation of night-train connections within Europe as a decision-making aid for climate-friendly labour mobility), Master thesis BOKU Vienna

DLR—Deutsches Zentrum für Luft und Raumfahrt, Luftverkehrsbericht (engl.: German Aerospace Centre, Air Traffic Report) (2012)

FLEXICOACH (2012) Final Report

Hödl (2006) Der europäische Markt für Nachtreisezugverkehr/Ein empirische Analyse der Nachfragedeterminanten (engl.: The European market for night train services/An empirical analysis of demand determinants), Schriftenreihe des Institut für Transportwirtschaft und Logistik WU Wien/ Publication Series of the Institute of Transport Economics and Logistics Business University Vienna, no 2

http://caboruivo.ch/2012/04/06/sparmassnahmen-keine-hochgeschwindigkeitszuge-in-portugal/ [17.06.2014]
http://de.wikipedia.org/wiki/Hochgeschwindigkeitsstrecke [17.10.2018]
http://de.wikipedia.org/wiki/Kupplung_%28Railway%29#Automatic_Center_Buffer_Couplings [13.03.2014]
http://en.wikipedia.org/wiki/China_Railways_CRH1 [15.03.2014]
http://europa.eu/legislation_summaries/transport/rail_transport/l24015_en.htm [13.03.2014]
http://europa.eu/legislation_summaries/transport/rail_transport/tr0009_en.htm [14.05.2014]
http://europa.eu/rapid/press-release_MEMO-13-897_en.htm [04/10/2014]
http://www.bav.admin.ch/themen/02783/02788/index.html?lang=en [14.05.2014]
http://www.bbt-se.com/projekt/fragen-antworten/ [17.06.2014]
http://www.femern.de/servicemenu/faq/bedarf-wachstum-und-entwicklung [17.06.2014]
http://www.frankfurt-airport.de/flugplan/airportcity?pax&sprache=de&ext=/de/ [01.07.2014]
http://www.luftlinie.org/ [17.10.2018]
http://www.railway-technology.com/news/newsfrance-italy-sign-agreement-for-lyon–turin-high-speed-rail-line [17.06.2014]
http://www.serbianrailways.com/system/en/home/newsplus/viewsingle/_params/newsplus_news_id/41325.html [30.06.2014]
http://www.uic.org/com/article/forthcoming-inauguration-of?page=thickbox_enews [17.06.2014]
https://en.wikipedia.org/wiki/List_of_the_busiest_airports_in_Europe [17.10.2018]
https://www.oebb.at/ [17.10.2018]
Masuch (2014) Potentiale für Hochgeschwindigkeitsnachtzugverkehre (engl.: Potential opportunities for high-speed night trains), St. Pölten
Plank (2007) Dimensionierung von Gepäckablagen in Reisezügen (engl.: Dimensioning of luggage racks in passenger trains), diploma thesis, TU Vienna
Rambausek (2009) Netzgestaltung und räumliche Wirkung von Hochgeschwindigkeitsbahnnetzen im europäischen Vergleich (engl.: Network design and spatial impact of high-speed rail networks in a European comparison), Diploma thesis TU Vienna Faculty of Architecture and Spatial Planning
Rüger (2004) Reisegepäck im Eisenbahnverkehr (engl: Luggage in rail traffic), dissertation, TU Vienna
UIC Study Night Trains 2.0 (2013)
White paper (2011) Roadmap to a single European transport area—towards a competitive and resource efficient transport system

The Next Generation of Rail Talent: What Are They Looking for in a Career?

Janene Piip

Abstract Technology and automation are impacting the rail industry with the requirements for interoperability across borders and systems. New skills are required for employees such as the ability to use contemporary communication technology, artificial intelligence and robotics. Without the skills of talented humans, it will not be possible to successfully transition to Industry 4.0. To prepare for a future rail workforce, talent practice and skill development now must consider the needs of Millennials who will be the rail workforce of the future. The talent process needs to be actively incorporated as a living and dynamic process by HR departments into business processes for strategic excellence. Organisations that have successfully integrated exemplary talent practice to propel their business are characterised by actions around proactive talent management, planning, diversity and inclusiveness and development of individual careers.

Keywords Rail transport · Skills · Rail career · Expectations · Rail workforce

1 Introduction

The problem of how to attract young talented professionals to rail organisations in an environment of increasing competition and choice is a real and ongoing challenge for the global rail industry. This topic has been on the workforce agenda for at least ten years and after a decade, is still widely debated. With plenty of discussion about attracting and retaining talent, skills shortages, mismatches of skills for new jobs and jobs disappearing due to artificial intelligence and the increasing use of robots, few rail organisations acknowledge that the 21st century is a person-centred age. Many of the workforce issues organisations cite about not finding the 'right' people or skills mix result from failure to communicate the right value proposition to the right people through the right channels. Once engaged, developing and retaining staff is all about doing the right things at the right time through customised support for

J. Piip (✉)
JP Research & Consulting, PO Box 2614, Port Lincoln, SA 5606, Australia
e-mail: janene.piip@gmail.com

© Springer Nature Switzerland AG 2020
M. Marinov and J. Piip (eds.), *Sustainable Rail Transport*, Lecture Notes in Mobility,
https://doi.org/10.1007/978-3-030-19519-9_10

each person. Sustainable rail talent practice is, therefore, about leadership, people and communication (Short et al. 2011). As Robert Bolton described in his seminal text 'People Skills' (1987), 80% of the problems in the workplace are related to communication ....

In this scenario lies the essence of this paper which focuses on attracting talented staff to rail organisations who will be the workforce of the future. It explores what needs to be considered by rail organisations wanting to make their organisation more attractive to young professionals. These are young people starting in their career, broadly described as those around the age of 30 years and under, or the age group as described as 'Millennial' (Dimock 2018). Drawing on the literature and findings from an empirical study conducted with 75 young professionals about career values and attitudes (Piip and Jain 2015), this work has valuable lessons for rail organisations as they contemplate their future workforces.

2 Background

With the future is upon us, talented people will facilitate the next workplace iteration. Industry 4.0 or the next Industrial Revolution uses 'transformative technologies to connect the physical world with the digital world' (Department of Industry 2018). Traditional business challenges, such as labour costs, will be changed dramatically through robots and artificial intelligence replacing low skilled and repetitive work. Talented professionals with the right technical skills, problem solving abilities and well-developed resource management capabilities will be highly sought after to lead and drive business productivity and efficiency. In the light of these trends, these new person-centred currencies required for sustained commercial achievement in a knowledge based economy (Agarwal and Green 2011), highlight projected rail workforce difficulties (Wallace et al. 2010). Rail organisations will have to widen the pool of people to fill these roles, by considering all genders and demographics.

In Australia, the changing demographics of the Australian population have further highlighted the reducing potential rail labour supply in the next 30 years (Productivity Commission 2005). The findings reveal that the total workforce will comprise older as well as younger workers (Millennials) with a shrinking middle-age sector. Since the year 2000, there has been a marked decline in the birth rate and the potential pool of workers is slowly diminishing. By 2045, the largest age group of potential employees will be between 45 and 55 years of age with a decline in the contribution of older people to work. The total number of older people working will not meet the growing needs of the rail industry, especially in new job areas with increasing skill needs.

Given that the front line is often the entry point to an organisation, organisational leaders should be recruiting people actively from diverse backgrounds and genders to remedy these impending scenarios. Considering these points, four key challenges facing rail organisations include:

- Understanding the critical jobs in the organisation
- Attracting and recruiting talent for those roles
- Training and Developing talent
- Retaining talent.

2.1 The Reality of Talent Practice

To start the discussion, I offer two examples from personal experience as a talent management consultant, writer, educator and career practitioner. In the last five years, hundreds of people have been assisted to find their 'perfect' match between their skills and the needs of employers, facilitating the start of career fulfilment for many individuals. The examples relate to stories from my own as well as the young professional client's lived experiences, encompassing what happens when employers are looking for talented staff, and young professionals are gaining a career start. In the first example, a client, a talented rail engineer with both workplace experience as well as a relevant doctorate qualification was hoping to gain a role in a reputable rail organisation. In his efforts to find employment, he had applied for many positions, but was not successful with any of them, describing his greatest difficulty as knowing what the company was really looking for in an employee. Even with all the relevant qualifications according to the job description, he did not secure any of the roles. He requested a tool to: '*Help him assess his own skills and match them to jobs that they are required in*'.

The learning from this experience highlighted that job descriptions developed by rail companies, in their efforts to attract the most appropriately qualified staff, are not often well-defined. While they usually describe the technical requirements such as qualifications and what the role might entail, they leave out considerable detail about the organisation, the organisational strategic intent, behavioural competencies, the organisation's values, the culture and the all-encompassing skills required for the position. For example, they often fail to outline how much time or percentage areas of work will require, the Key Performance Indicators (KPIs) of the position, the digital and soft skills required for the role or the critical challenges of the position. Young professional candidates are seeking more detail in assessing employment opportunities so that they can see how their skills fit into the organisation. They want to minimise their time wastage by applying for the 'right' positions, especially when this may be their first role in the workforce.

The next example offered is about three clients who received coaching for one organisational vacancy. With knowledge of the job specification and role, as well as knowledge of the individuals' skills and experiences, candidates were assisted in preparing their applications with the objective of obtaining an interview. In time, all clients underwent an interview with a panel of three staff. At the meeting, questions for each applicant varied, and the panel had varying views of the role requirements. In summary, the person with the most appropriate skills for the role, regarding personality, skills and cultural fit was not selected. Instead, the panel recruited a candidate with a good interview performance. However, this person did not fit the job description requirements or the style or culture of the organisation.

The lessons to be learnt from these examples point to the need for detailed and documented role requirements and a shared understanding of the skills, talents and abilities the new employee should demonstrate, otherwise the wrong person will be selected. There are organisations filled with staff chosen by practices in the examples quoted here. Selection based on 'gut feel' or true analysis of person-fit with the organisational culture, potential or employee's personality is not the best practice to recruit the next generation of talent.

2.2 The Employer of Choice Concept

As all rail companies understand, innovation in business is now a vital source of competitive advantage, requiring regular examination of existing practice, the development of new ideas and consideration of how best to manage resources. This statement is particularly salient in light of this discussion that seeks to identify some of the main factors that contribute to attraction and retention of new talent. To help one learn and grow, looking to practices in other leading sectors provides opportunities for the rail industry to emulate practices undertaken elsewhere.

The Employer of Choice (EOC) concept taps into attracting and retaining talent in a global and uncertain, ever-changing economy. The EOC concept is identified in recent workforce development projects including those within the Australian rail industry (Pricewaterhouse Coopers 2006), in Local Government (Tmp Worldwide 2006) and the finance sector.

'Employer of Choice' is a concept that considers the views and perspectives of organisations, managers and employees to provide a holistic view of whether the company is a good place to work. Common themes linking Employer of Choice businesses relate to the high value given to talented staff and the ability of businesses to engage and retain their employees for some years (Gill 2008). Broadly speaking, the Employer of Choice concept 'means being a business or workplace that aligns the changing needs and interests of the individual and the organisation' (Baker, 2014). In explaining the Employer of Choice concept, identified drivers and characteristics in becoming an 'Employer of Choice' are outlined (Table 1).

2.3 Life Stages, Young People and Careers

The time of career commencement, from 18 up to 30 years, has long been considered in adult development writing, as broadly a time of exploration and self-discovery (Kegan 1998; Kegan and Lahey 2010). These years are characterised in the literature by features such as investigation, instability, self-focus, feeling in between and new possibilities (Arnett 2015). As a young person moves from school to university to

Table 1 Three models of Employer of Choice

Model 1 (Hull and Read 2003)	Model 2 (IBM Business Consulting Services 2005)	Model 3 (Baker 2014)
Fifteen key drivers of an Employer of Choice: • the quality of working relationships • workplace leadership • having a say • clear values • being safe • quality environment • recruitment, pay and conditions • getting feedback • autonomy and uniqueness • a sense of • ownership and identity • learning • passion • having fun • community connections	Successful organisations: • source and manage talent in line with their business strategy • hold leaders accountable for the development and growth of people • utilise organisational knowledge effectively • Manage human resources (HR) like a business, linking HR results to business outcomes	In a new economy, the psychological contract between employer and employee is characterised by: • Flexible deployment • Customer focus • Performance focus • Project-based work • Human spirit and work • Commitment • Learning and development • Open information

work, many individuals are still deciding on their real career interests, their values and whether they have completed the right course of study. It is unrealistic to expect embedded, lifelong careers to be decided until after 30 years of age because there are so many possibilities for young people to explore in these formative adult years. Flexibility to explore various options has, and always will be, a characteristic of this life stage.

This stage of self-exploration or the reasons why embedding a career in the rail industry at this stage is troublesome for many, is scarcely mentioned in the rail workforce development debate. Instead, the focus seems to be more about the fact that the rail industry is not as attractive to young talented professionals as other industries (Wallace et al. 2010).

In rail organisations of past times, young people left school and gained employment with little knowledge of the employing company. However, society has changed dramatically over the last 40 years, and goals and aspirations are different. The internet, technology and social media have provided the opportunity to access information about careers and organisations in ways never imagined in the 1970s. In 2018 and beyond, young people use social media to research companies, participate in online forums to explore the culture of the company and consider the comments from existing employees about the workplace. Having grown up in a digital world, many young professionals are familiar with the internet, social media and finding relevant information online. They expect technology to enhance their job and expect organisational processes to be technologically connected and instant

(Workfast 2016). These realities have substantial implications for rail organisations in the search for the best talent.

2.4   Looking to Talent Practices in Other Organisations

In considering these points, we looked to LinkedIn, a social networking site, where professionals can develop networks and connections, and post their workplace skills and experiences online. Companies seeking to employ professionals with specific skill sets can search the online LinkedIn database for a suitable candidate who may meet their staffing needs. We visited LinkedIn, as Wallace et al. (2014) had done previously to find out about which organisations were leading the field in 2018, according to users of the site, regarding some workplace practices valued by potential employees.

LinkedIn conducts a poll of the most sought-after companies to work for each year, according to four factors including how well the company is known, how much audience interaction there is with the posted content such as articles, how much interest the company's jobs advertisement generates, and how well the company retains professionals employed through LinkedIn connections. In 2018, the companies rated most highly by the LinkedIn audience included (Lobosco, 2018):

1. PwC Australia—professional services
2. Commonwealth Bank—finance and banking
3. Deloitte Australia—audit, consulting, financial advisory, risk management, and tax services
4. KPMG Australia—professional services and auditing
5. Westpac Group—finance and banking
6. CIMIC Group Limited—engineering-led construction, mining, and public, private partnerships
7. Macquarie Group—global diversified finance company
8. Lendlease—developer of public, private housing for defence
9. National Australia Bank—finance and banking
10. Ernst and Young (EY)—professional services and accounting.

Since the work completed by Wallace et al. (2014), almost five years ago, the list has changed dramatically. In 2018, global companies in financial and professional services and banking dominate the list, whereas, in 2014, mining, construction, utilities and transport featured highly. These findings reflect the changing Australian economy, but they also reflect young professionals' desire to work in global companies with the availability of diverse career opportunities and the opportunity to travel. While there are no rail companies are on this list, the lessons to be learnt here point to the need to market vacant positions using the right channels where the target group of professionals are searching for opportunities. Beyond initial marketing and connection, other factors contribute to brand awareness and industry leadership outlined in the next section.

2.5 What Millennials Want in a Career

Of current concern to young people is that of business ethics in companies where they are considering developing their career (Deloitte Touche Tohmatsu Limited 2018). Chalofsky and Krishna (2009) write about young professionals' desire to work for socially responsible and ethically driven organisations because on their journey of self-discovery, young people want to find meaning in their work and examine how espoused company values align with their personal principles. Doing the right thing is an important concern for young people in working toward their career goals and providing personal direction for future opportunities.

Many rail companies are multinational corporations. Young Millennial professionals, who have grown up in an era of social change, believed that large companies have the potential to be instrumental in job creation for the less fortunate (such as long-term unemployed, marginalised groups, culturally and gender diverse people). They have the ability to address discrimination and inequity and be socially responsible by improving the lives of others (Deloitte Touche Tohmatsu Limited 2018). Many are easily disillusioned if they discover that the company publicity does not match with real and experienced workplace practices.

3 Our Study About Career Values and Attitudes

As mentioned previously, a global study we conducted about rail professionals attitudes to their careers (Piip and Jain 2015) highlighted some key findings of careers for the under 30 year age group or those in the Millennial age range. The study was conducted using an online survey tool and distributed to rail professionals in 30 countries through email notification to employer organisations and LinkedIn promotion. While the study was intended for professionals in the mid-term of their career, 75 young professionals under 30 years of age responded to the survey. The results of the younger age group have been segregated to develop this paper.

The next section of this paper explores in more detail, findings from the literature and our study about five themes including leadership, wellbeing, work-life balance, career advancement and learning that influence the attraction and retention of young, talented professionals to rail organisations.

3.1 The Role of Leadership in the Attraction and Retention of Talented Professionals

Many writers recognise that leadership is a complex skill set, made even more complicated due to the external world becoming increasingly ambiguous and uncertain (Short et al. 2011). For leaders, it is often a step into the unknown. Leadership has

Table 2 The difference between management and leadership

Management *Produces order and consistency*	Leadership *Produces change and movement*
Planning and budgeting – Establish agendas – Set timetables – Allocate resources	Establishing direction – Create a vision – Clarify the big picture – Set strategies
Organising and staffing – Provide structure – Employee people – Establish rules and procedures	Aligning people – Communicate goals – Seek commitment – Build teams and coalitions
Controlling and problem-solving – Develop incentives – Generate creative solutions – Take corrective action	Motivating and inspiring – Inspire and energise – Empower and enable – Satisfy unmet needs

Source Northouse (2010)

an essential role in attracting and retaining talented professionals, especially those starting their careers. Leadership sets the tone of the workplace and sends messages to the new employee as to whether it is a safe place to work (regarding discrimination, harassment or inequity), whether the organisation can fulfil career dreams or whether it is a place to escape from as soon as possible. The decision to stay or leave the organisation is regularly decided during the induction process and early days of employment.

Factors that contribute to good leadership clearly point to skills that require the leader to engage with people by setting a vision or direction, communicating and clarifying that direction, and then selecting, empowering and enabling talented people to be a part of that vision. Managers should look after the new employee to ensure they settle in well. Unfortunately, many rail professionals are appointed to a management role in a company but soon realise that not only do they have to manage physical resources, but the role entails guiding, leading and empowering staff. Many require further training to do this job well.

Table 2 outlines, for consideration, the difference between management and leadership highlighting that many skills of leadership are 'soft' or relationship skills that can be developed further and learnt over time with the right interventions.

All background discussion aside, Snee and Hoerl (2012) suggest that good leadership (change management, improving work and looking after people) should occur around 70% of a leader's time while management (maintaining order and consistency, doing work) is estimated to take around 30% of ones' time.

For Millennials, leadership helps and empowers them to find their place within rail companies. Our survey showed that young professionals want leaders to guide them and help them develop their career as demonstrated in this comment:

I would like more opportunities for self-realisation (Rail Professional with MBA, under 30 years)

Leaders have a role in demonstrating their ability to be strong role models so that younger professionals can emulate their style and actions. In individual companies, higher level leaders need to consider how their leaders—both peers and at lower levels - are viewed by others and whether they have a positive standing or rating. Leaders convey the culture of the organisation, the ethics and the support structure to enable talented employees to succeed. They provide examples of the traits that are valued by the organisation—either good or bad. Many difficulties experienced by rail organisations in getting young people to eventually fulfil leadership roles relate to poor leader role models throughout their careers and the observation that leadership is a demanding role as highlighted in this comment by a manager in an older age group:

> Really it is poor managers and leaders who have no idea how to leverage my talents and the talents of other highly skilled and competent professionals (Human Factors Specialist)

He emphasised that talented employees want supportive and authentic leaders who can communicate appropriately and support younger professionals.

> They are bent on advancement and office politics at the expense of developing the skills and capabilities of staff in a modern rail business… leadership is …not being stuck behind a desk staring at a computer (Human Factors Specialist)

The skills and abilities needed for rail leaders are critical for organisations in transitioning change and encouraging the next generation of employees. Current rail leaders, especially at the frontline and entry-level roles where new professionals commence their career, have often transitioned from being a technical expert to overseeing people. Understandably, they are more comfortable with specified solutions and outcomes, and many leaders require assistance with the 'people-side of the business' (Piip 2014). Greater understanding of the diverse range of capabilities required for leadership in rail organisations today would enable stronger connectedness of leaders with the organisational purpose, and the potential to guide and develop both current and new employees. The sustainable goal for organisations is to have leaders who become more humane, moving from 'cop to coach' (Mcmahon 1993). In an environment of change, leaders, therefore, need more development so they understand the influence they can have on their employees and their ability to grow potential in others for the future.

3.2 Wellbeing

Leaders and managers should be concerned about the welfare of their staff, especially young professionals starting out in their careers. Keeping up with the demands, expectations and consequences of modern life affects mental health and wellbeing both socially and in the workplace. 'Keeping too many tabs open will drain your battery' is a convenient euphemism for modern life and especially for Millennials who are the most tech-savvy generation (Vogue 2017). For many young professionals, it is often difficult to draw the boundaries on work, play and rest when some are still developing as adults. Leaders and managers have a role to play in positively influencing young professionals through their leadership style to ensure a happy and productive working environment free from discrimination, bullying and the monitoring of unrealistic expectations. In our study, young professionals listed the three most significant challenges they faced in their career as:

1. Career Advancement 36.2%
2. Work/life balance over 30%
3. Talent recognition 22.4%.

These issues affect positive wellbeing and mental health as Millennials try to find their place at work.

3.3 Work/Life Balance

Deloitte (2018) report that flexible work arrangements with accommodating workplaces with are important to Millennials. Organisations that appear to have rigid structures such as rail companies need to rethink how they will engage and retain Millennial workforces in the increasing competition for talented professionals. Many young people see the gig economy as appealing compared to more formal work arrangements as it allows work flexibility, the opportunity to work from any location, anywhere in the world and undertake work assignments that have a beginning and an end.

In our study, respondents discussed issues of managing time for work and family when their company had essential deadlines to meet. Starting ones' career younger rail professionals found managing time for workplace projects challenging as demonstrated in this comment:

> There are many challenges that you must face at a workplace, especially when there are deadlines to submit your financial reports and there are too many other tasks you should do. When you do not have enough time in a work day, you will have to work after 6.30 pm until you finish all your tasks. And then you decide how to plan and manage your own work schedule so that you will make progress with all the tasks in your schedule and go home in time.

3.4 Career Advancement

Younger staff up to 30 years of age cited being able to advance their careers as a real issue while some also listed talent recognition as a critical factor:

> Create more opportunities for career advancement (Graduate, Subject Matter Expert)

In our complete survey (Piip and Jain 2015), the career issues for all age groups related to recognition of talent and career advancement, making recognition of talent almost 70% of all challenges faced by all employees. The message is that professionals do not feel like companies are using their skills or potential! When asked in our survey about what a young rail professional would do about the situation with his career, he answered:

Use Google Help (Employee without Management function, Lead engineer)

3.5 Career Development

Once engaged in a career role, companies really have the years from 18 to 45 to tap into employee enthusiasm to develop their career within that company (Piip and Jain 2017). By 45 years, employees' enthusiasm for their career tends to stabilise if they have found their career niche. An upward trajectory can be identified for those who step into a management role while those who become specialists or experts tend to develop broad skills in their chosen field. Companies are at risk of losing Millennials if they do not maximise individual potential by offering opportunities for career development, or show interest in their career progression, at this stage.

3.6 Career Conversations

Having a career conversation with employees is the most critical job of leaders so they can find out whether the individual is engaged with their career or whether they want to develop in a different direction. Communication is the key to identifying career development needs, career goals and aspirations. As this young professional described:

In order to help an individual to build his professional skills, organisations should keep up with new ways of interaction between employees and managers using the latest technologies (Subject Matter Expert, Railway Association)

Career conversations should be scheduled on an ongoing basis and built into everyday occurrences rather than scheduling a meeting once a year. Career conversations help employees to see a future for themselves in the business and give them time to think through the feedback and interactions with their leader or manager. These discussions all take time but should be on the forefront of every leaders' mind and should not be incorporated with performance discussions. If career conversations are incorporated with performance discussions, this sends mixed messages to the employee about whether the company is trying to reward or punish them. Performance discussions should be held as soon as there is an issue with performance while career conversations should be held frequently and on an ongoing basis.

Table 3 Learning opportunities valued by Millennials

60.2% of Millennials want opportunities to participate in exchange programs to other countries
I think that the company must have a special exchange program to other companies and other countries (Employee with a Doctorate, without a management function)

49.6% of Millennials want to undertake different roles through job rotation
Offer Opportunities to develop within the organisation (Employee without management function)

46% of Millennials want **Workplace Experiences**
Have committees which include a broad range of experience and knowledge (Employee without a management function)

37.2% of Millennials want to **participate in conferences**, but many believed they were not considered to develop their skills through this process
First organisations need to accumulate money in the budget to give opportunities for individuals to develop their career and professional skills (Subject Matter Expert in a supplier to rail company)

31.9% of Millennials want to **research** their area of expertise
To provide an opportunity to develop the professional quality of specialists (Chief Specialist production and technical department)

29.9% of Millennials want to **undertake exchange within other companies**
Offer new and different development opportunities in other railway companies (Consultant to a rail company)

17.7% of Millennials want to **provide consultancy services to clients**
Less reluctance to promote and give opportunities to younger engineers (Consultant engineer to rail company)

14.2% of Millennials want **flexibility at work or part-time work** to develop career skills
allow goal-oriented worktime (Graduate, Freight Traffic)

12.4% of Millennials want to remain employed but **take time for a sabbatical**
They can manage talent and improve their skills by allowing them to study, giving them opportunity to increase their professional skills and rotate talented individuals in order to motivation. (Project Manager, Railway Association)

3.7   Learning for a Rail Career

We asked young rail professionals in our survey how organisations can help them to develop their professional and career skills, so they stay engaged in the industry, and they described that learning throughout their career is important to them (Table 3).

The comments from rail professional about the learning opportunities they valued provide insight into ways companies can keep this group engaged. Opportunities for career development, learning, self-exploration and talent recognition are enhanced by aspirations such as exchange to other countries, being the highest rating category of 60%. Millennials who are supported to learn and develop their career are more likely to stay with the organisation, expand their knowledge and feel valued in the work they do. Through ongoing learning, they can expand their knowledge and expertise as they are exposed to new ideas and perspectives, especially when they participate in activities such as job rotation, work committees, consulting work, conferences and networking activities. These activities take time to consider but should be an active part of a young professionals' career development plan.

4 Recommendations

To keep attracting and retaining young professionals to the rail industry, feedback from our survey participants highlighted points in critical areas that will raise awareness of the rail industry as a positive place to work or an 'Employer of Choice'. These recommendations include:

Hiring and job advertisements such as the use of social media and visual or infographic job descriptions to attract millennials are more appealing than tradition text-based descriptions. Use social media links such as LinkedIn on the job advertisement to reinforce and find connections with people already working in the company. Develop a recruiting strategy with job descriptions that read well on a smartphone as Millennials are used to reading on their mobile devices.

The use of images, videos or group of photographs can depict a company culture where young professionals would like to work. Dynamic job descriptions and approaches to recruiting portray the company as clean, green and flexible and don't require lots of printing. Current projects, employee testimonials and socially responsible activities can be linked to the organisational website to market the brand to showcase the organisation to potential talented staff.

Induction is a way for an employee to feel part of the new workplace with information that will enable the achievement of productivity within the role in a short time. An induction should include a workplace orientation, information on safety and job requirements, reporting lines, dress code, hours of work, and policies such as the use of social media at work.

Sustainable induction into a new work place is a process that happens over several months to assist the new professional grasp the information provided and find their way in the organisation. The establishment of a mentor or buddy system allows the opportunity to ask questions of different people throughout this process and to grasp new information in timely ways (Piip 2019). The impressions developed about the workplace by the new employee during the induction process will help determine whether they consider the organisation is an 'Employer of Choice'. The Induction process can be made interesting, welcoming and dynamic with some thought given to how the new employee will be welcomed and motivated to stay with the organisation for the long term.

Leadership plays a significant role in the induction process and throughout the working lifespan of a new professional. Millennials valued wellbeing, work-life balance, career advancement and career development. Traditional rail leaders, especially in frontline roles are skilled in technical areas and need more training to develop the softer skills of leadership that Millennials desire.

Learning and development opportunities requested by Millennials comprise activities that include work exchange to other countries and companies, conferences, job rotation, participating in research and consultancy opportunities and undertaking a sabbatical for recreation or further learning. Many of these opportunities do not require the cost of cost fees and could be arranged with well thought through planning.

Career development opportunities are valued highly by Millennials which require ongoing career conversations to identify the right learning experiences.

5   Conclusion

The next workforce revolution is characterised by the human age where rail organisations must find out what people want out of their career. The paper shed light on the factors that are important in attracting and retaining young professionals in the rail industry of the future. The importance of leadership, people and communication should not be underestimated in defining and identifying the skills and competencies needed for job roles across the entire organisation. Having a plan and identifying the jobs of strategic importance is, therefore, important in the quest for sustainable talent management.

The research study examined the views of young Millennial professionals in a survey about career values and attitudes to identify that career conversations and advancement were one of the most essential factors engaging them with their current role. Being valued, receiving feedback about their work and gaining recognition for work done well were all important and could be achieved through career development opportunities within the workplace. It was found that leaders had an important part to play in facilitating these learning opportunities, but many did not have the knowledge or experience to develop these activities. It is concluded that more effort is needed to assist leaders to develop the soft skills of leadership and learning and for Millennials to be included in more active learning such as learning, travelling and networking to engage them in their career.

Consequently, in conclusion, do you understand what your talented employees want? If not, it is time to communicate and have a conversation.

References

Agarwal R, Green R (2011) The role of educational skills in Australian management practice and productivity. In: Curtin P, Stanwick J, Beddie F (eds) *Fostering enterprise: the innovation and skills nexus—research readings*. Adelaide, NCVER

Arnett (2015) Emerging adulthood. Oxford University Press, New York

Baker T (2014) Attracting and retaining talent: becoming an Employer of Choice. Palgrave Macmillan, Basingstoke, Hampshire, UK

Chalofsky N, Krishna V (2009) Meaningfulness, commitment and engagement: the intersection of a deeper level of intrinsic motivation. Adv Dev Hum Resour 11:189–203

Deloitte Touche Tohmatsu Limited (2018) Deloitte millennial survey. Australia

Department of Industry, I. a. S (2018) Industry 4.0 [Online]. Australia: DIIS. Available: https://www.industry.gov.au/funding-and-incentives/manufacturing/industry-40

Dimock M (2018) Defining generations: where Millennials end and post-Millennials begin [Online]. Pew Research, USA. Available: http://www.pewresearch.org/fact-tank/2018/03/01/defining-generations-where-millennials-end-and-post-millennials-begin/

Gill R (2008) Reputation and Employer of Choice for Australian business. Monash Bus Rev 4

Hull D, Read V (2003) Simply the best workplaces in Australia—working paper. Australian Centre for Industrial Relations Research and Training (ACIRRT), University of Sydney, Sydney

IBM Business Consulting Services (2005) The capability within – the global human capital study 2005. IBM Corporation, United Kingdom

Kegan R (1998) In over our heads: the mental demands of modern life. Harvard University Press, Massachusetts, USA

Kegan R, Lahey LL (2010) Adult development and organisational leadership. In: Nohria N, Khurana R (eds) Handbook of leadership theory and practice: an Harvard Business School (HBS) centennial colloquium on advancing leadership. Harvard Business Press, Boston, MA

Lobosco M (2018) Here are the top companies of 2018 [Online]. LinkedIn. Available: https://business.linkedin.com/talent-solutions/blog/employer-brand/2018/here-are-the-top-companies-of-2018. Accessed 27 Nov 2018

Mcmahon F (1993) From cop to coach: the shop floor supervisor of the 1990s. In: Caldwell B, Carter E (eds) The return of the mentor: strategies for workplace learning. The Falmer Press, London, UK

Northouse PG (2010) Leadership: theory and practice. Sage Publications Inc, Thousand Oaks, CA

Piip J (2014) Exploring leadership talent practices in the Australian rail industry. Doctor of Philosophy, University of South Australia

Piip J (2019) Mentoring for career development: organisational approaches to engage and retain employees. In: Fraszczyk A, Marinov M (eds) Sustainable rail transport. Lecture notes in mobility. Springer, Cham

Piip J, Jain A (2015) Talent management of experienced professionals in the rail industry. Australia and Switzerland

Piip J, Jain A (2017) Mid-career rail professionals: new approaches for companies to engage employees and capture knowledge. In: 4th UIC world congress on rail training, Potsdam, Berlin

Pricewaterhouse Coopers (2006) The changing face of rail: a journey to the employer of choice. Australasian Railway Association (ARA) Inc, Canberra, ACT

Productivity Commission (2005) Economic implications of an ageing Australia [Online]. Productivity Commission, Melbourne. Available: http://www.pc.gov.au/projects/study/ageing/docs/finalreport

Snee RD, Hoerl RW (2012) Leadership—Essential for developing the discipline of statistical engineering. Qual Eng 24(2):162–170

Short TW, Piip JK, Stehlik T, Becker K (2011) A capability framework for rail leadership and management development. A research report produced by the CRC for Rail Innovation (May 2011). Brisbane, Australia

Tmp Worldwide (2006) Promoting Local Government in South Australia as an employer of choice—report of findings, insights and recommendations [Online]. South Australia Local Government Managers, Adelaide. Available: http://www.lga.sa.gov.au/webdata/resources/files/Promoting_Local_Government_in_South_Australia_as_an_Employer_of_Choice.pdf. Accessed 18 July 2014

Vogue (2017) Millennials are the most tech-savvy generation in human history, and the most anxious. Coincidence? [Online]. NewsCorp, Australia. Accessed https://www.vogue.com.au/beauty/wellbeing/why-millennials-are-the-most-anxious-generation-in-history/news-story/755e7b197bdb20c42b1c11d7f48525cd 2018

Wallace M, Sheldon N, Cameron R, Lings I (2014) What do young Australian engineers want? Strategies to attract this talent to less glamourous industries. In: Short T, Harris R (eds) Workforce development—strategies and practices. Springer, Singapore

Wallace M, Sheldon N, Lings I, Cameron R (2010) Attraction and image for the Australian rail industry. In: British Academy of Management conference. University of Sheffield, Sheffield, UK

Workfast (2016) The wave of human capital: recruiting millennials [Online]. Australia. Available: https://workfast.com.au/blog/recruiting-millennials/

Professional Rail Freight and Logistics Training Programme: A Case Study of Energy and Petrochemical Company in Bangkok, Thailand

Kaushik Mysore, Mayurachat Watcharejyothin and Marin Marinov

Abstract An intensive training course on rail freight and logistics was delivered in English to employees from the energy and petrochemical industry in Thailand in 2017. It was organized with the purpose of helping their staff gain knowledge and understanding of principles for effectively managing rail freight and logistics systems. This training course discussed the potential for economic growth and readiness of Thailand to become the logistics rail-based hub for ASEAN. Participants had less experience in rail freight and logistics; hence they wished to improve their knowledge of the subject area for potential management of rail freight and logistics projects across Thailand and its neighbouring countries. After the five-day training course, feedback from participants has been collected; the analysis showed positive views. The participants found the course helpful as it met their expectations. They also provided constructive criticism and useful recommendations for the future delivery of this course.

Keywords Rail freight · Logistics · Training · Feedback

1 Introduction

Rail freight network operations in Thailand were established in 1890 and have been progressively improved since then. However, during World War II Thailand encountered a financial crisis, and its railway network and operations were interrupted. The country changed its direction and chose a lower cost system, using roads. As a result, Thailand started to focus on using trucks to transport goods and passengers. Consequently, domestic freight in Thailand became truck-based holding an 80% market

K. Mysore
Queen Alexandra Sixth Form College, North Shields, UK

M. Watcharejyothin
School of Environment Resources and Development, Bangkok, Thailand

M. Marinov (✉)
Engineering Systems and Management, Aston University, Birmingham, UK
e-mail: m.marinov@aston.ac.uk

© Springer Nature Switzerland AG 2020
M. Marinov and J. Piip (eds.), *Sustainable Rail Transport*, Lecture Notes in Mobility,
https://doi.org/10.1007/978-3-030-19519-9_11

291

share in total and only four percent of its freight was transported by rail (NESDB 2016a, b).

Recently, the Thai government launched a new plan for the country's logistics strategy, aiming to become a "Logistics Hub of ASEAN" using the rail-based multi-modal logistics. A budget of 425,000 million Thai Baht is to be spent on a double rail tracks project and logistics facilities (OTP 2017a, b). It is anticipated that the investment will increase the country's rail freight capacity by five times of the current capacity before 2026.

Through this significant transition, Thailand faces many new challenges. In addition to sourcing hardware facilities, locomotives, bogies, and logistics equipment, Thailand needs to increase its manpower capacity and train people for efficient rail freight operations. Currently, there are 15,000 employees working for the State Railways of Thailand. It is estimated that an extension of rail freight development could involve around 150,000 staff employed in the rail freight and logistics industry in 2026. This estimation is calculated based on the number of staff working in the rail industry per rail distance in kilometres (50 staff per kilometre) and multiplied by railway track expansion from 300 to 3000 kilometres (SRT 2018). In Thailand, the ratio is much higher than the case of other well-developed railway systems in countries such as Germany (4.6 staff per kilometre), UK (10 staff per kilometre) and India (19.4 staff per kilometre) (Thairailtech 2017; Indian Railways 2017) because Thailand still uses a lot of labour-intensive railway technology (SET 2017).

With a foreseeable high demand of manpower in rail freight operations, Thailand may face a shortage of labour supply soon. Therefore, in seeking an alternative platform for staff training on rail freight and logistics, a five-day intensive training course has been organised and delivered to employees from an energy and petrochemical industry in Thailand. The course aimed at providing knowledge and understanding of basic principles and operations management of rail freight and logistics. A five-day intensive training course of rail freight transportation and logistics for staff in selected industries was held in Bangkok, Thailand for the first time during 22–24 November 2017. The course contents included freight fundamentals, rail economic management and planning, urban freight by rail, rail freight current challenges and prospective, freight and logistics services and rail freight interchanges. The rest of the paper is organized as follows: Section two provides examples of training initiatives for skills development and rail freight and logistics, Section three presents the methodology, discusses the feedback data and analyses the results and a discussion of key messages is presented in section four, followed by conclusions and recommendations in the last section.

2 Training Initiatives in Rail Fright and Logistics

Skills development in rail freight and logistics have been facilitated by university courses, intensive programmes and workshops. Readers can refer to the following publications outlined further in this section.

Several MSc programmes, developed to train potential rail workers, are available within Europe. These courses are based around rail freight and logistics, as well as infrastructure and railway systems engineering. They also offer some research modules which are almost like a research apprenticeship, helping students to learn in the rail work environment. University courses have been discussed by Marinov et al. (2013), Marinov and Fraszczyk (2014), Fraszczyk et al. (2016), Tsykhmistro et al. (2014), Lautala et al. (2011), Marinov and Fraszczyk (2013a, b). The rail freight and logistics is a booming sector which cannot be sustained without education and training. Global logistics have contributed significantly to the growth and revitalisation of rail freight. In combination, rail freight serves many of our social needs without causing severe damage to the environment (Woroniuk et al. 2013; Marinov et al. 2010a, b, 2011a, b; Marinov and Viegas 2011). Education through university courses helps improve the strategic and tactical management of rail freight and logistics service providers as it increases theoretical knowledge of personnel involved and their ability to solve complex operational problems.

Intensive programmes delivered over a short period are another tool for skills development. These programmes involve academic learning over a short period. They can be attended by undergraduate students. Since professors working with the rail industry are delivering these courses, the quality is likely to be high. Some of these programmes are discussed in Drobisher et al. (2016), Fraszczyk et al. (2012, 2015), Marinov and Ricci (2012).

Workshops provide another method for bridging the gap in any particular area of skills development in the rail freight and logistics industry. A workshop would typically include a series of lectures and discussions on various topics to enhance theoretical knowledge of attendees. Group exercises may then follow to test practical skills learned. Workshops are usually organised with the aim to discuss common problems faced in the whole industry. Workshop participants include employees, stakeholders, final year undergraduate students and graduates. The benefits of attending workshops include exposure to deferent experience, learned better practices, practical skills development and increased technical knowledge. Examples of organising and running successful workshops include Fraszczyk et al. (2012), Dawson et al. (2017), Fraszczyk et al. (2015a), Marinov and Fraszczyk (2013a, b).

3 Methodology

The staff from Energy and Petrochemical Industries (EPI) in Thailand who attended the training course handle products such as crude oil and petroleum petrochemical products in liquid tank and containers, using rail freight transport. Therefore EPI staff from the same organisation, with less experience in logistics were invited to attend the course.

On the last day of the training, course participants were provided with feedback forms (Fraszczyk et al. 2016) to investigate participants' views on the course. The feedback form consisted of 11 questions. It was designed to allow the participants

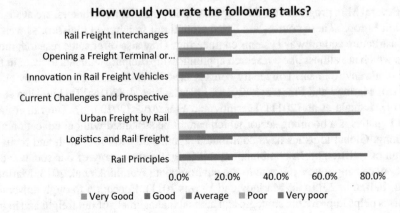

Fig. 1 Percentage of rating of talks

to include their thoughts and opinions about the course in a descriptive manner. Questions in the feedback form were quantitative, qualitative and a mixture of both. The rating system in the form ranged from Very poor/negative to Very good/positive. Data (answers) was processed into tables and graphs using Microsoft Excel for comparisons and analysis.

4 Questions and Answers

Question One: Rate talks

Participants were asked to rate seven talks (1. rail principles, 2. rail freight and logistics, 3. urban freight by rail, 4. current challenges and prospective, 5. innovation in rail freight vehicles, 6. opening a freight terminal, 7. rail freight interchanges), using the following scale: Very poor, Poor, Average, Good and Very Good. The lowest rating given for all the talks was Average. "Urban freight by rail" had the lowest rating: 25% of participants rated the talk as Average, and 75% participants rated the talk as Good. "Innovation in Rail freight vehicles" had the highest rating: 22% participants rated the talk as Very Good, 67% participants rated the talk as Good, and around 11% participants rated the talk as Average. Figure 1 is the graphical representation of the rating of talks by the participants. Urban freight by rail is not a priority area for the Thai government at the moment, which could explain why this talk was given the lowest rating. As for innovation in rail freight vehicles, the current rolling stock in Thailand is old and outdated. Hence the participants were very interested in updating their knowledge on rail freight vehicles and built up plans for what might be a suitable rolling stock for the Thai rail network in the near future.

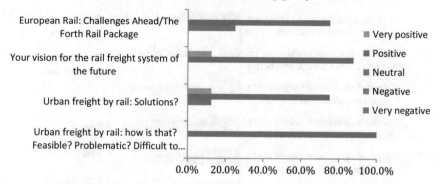

Fig. 2 Group exercise evaluation

Question Two: Rate group exercises and experiences

Participants were asked to rate specific group exercises (European Rail, Urban Freight by Rail—how is that?,—Solutions?, Rail freight system of the future) and their experience of those exercises. The rating was done using a "one to five" scale. The lowest rating given for any group exercise and experience was *three*. The group exercise on "European Rail: Challenges Ahead/The Fourth Rail Package" had the lowest rating: 25% of participants rated this exercise as Neutral, and 75% of participants rated the exercise as Positive. "Your vision for the rail freight system of the future" had the highest rating: 87.5% participants rated this exercise as Positive, and 12.5% participants rated the exercise as Very Positive. Figure 2 shows the graphical representation of the rating of group exercises. Although the participants found the group exercise on "European rail—forth package" quite interesting, this group exercise was rated the lowest. It is because the political rail framework in Europe is not suitable for the current regulatory system in Thailand, and as a result, this group exercise was not given a serious thought.

On the contrary, due to its nature, timeliness and importance for rail freight developments in Thailand, the group excercise on "...rail freight system of the future" was given the highest rating.

For the experiences, "Group discussions helped me improve communication skills" had the lowest rating: 50% participants rated this experience as Neutral, 37.5% participants rated the experience as Positive, and just 12.5% participants rated the experience as Very Positive. Group discussions "improved knowledge in rail freight and logistics" and "happy with the support from other participants" had the highest rating: 25% participants rated the experiences as Very Positive, 62.5% participants rated the experiences as Positive, and 12.5% participants rated these experiences as Neutral. Figure 3 shows the rating of the graphical representation of experience, followed by Figure 4 which compares group excersise and experience.

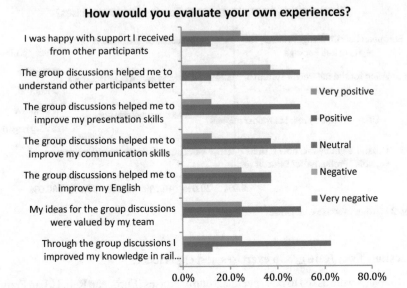

Fig. 3 Experience evaluated

Question Three: Favourite talk(s) with reasons

This was the first descriptive question asking the participants about their favourite talk/topic along with a reason to support their answer. Participants indicated they liked these talks/topics the most with: "Rail freight interchanges", "Vision for the rail freight system" and "Rail freight system in the UK" rating highly. Even though the participants liked the same talk/topic the reasons given were entirely different and unique. This information is illustrated in Table 1. Three participants did not respond to this question.

Question Four: Language barrier—Yes or No

Participants were asked if English language was a barrier in understanding the talks. Furthermore, they were asked to support their answer with a reason. The majority of the participants voted No—55.6% while the rest voted Yes—44.4%. Figure 5 shows the graphical representation of participants who voted Yes and No.

37.5% of participants voted No due to unknown technical terms. 12.5% of participants voted No due to unclear instructions and lack of railway background. The rest had unique reasons for voting their respective options. Figure 6 shows the graphical percentage distribution of participants' reasons for voting Yes or No.

Question Five: Recommend course to others—Yes or No

Participants were asked if they would recommend this training course to other delegates. Similar to question five, they were to choose Yes or No and support their answer with a reason. 71.4% of participants voted Yes while the remaining 28.6%

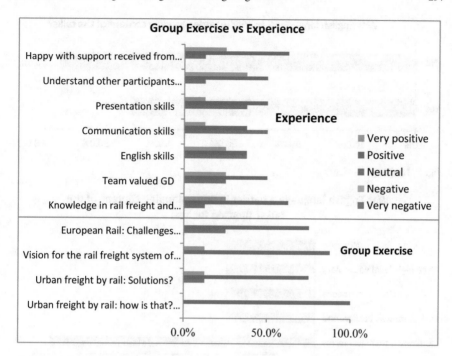

Fig. 4 Comparison of group exercise and experience

Table 1 Favourite talk/topic and why

	Talk/topic	Reason
	Which talk/topic on Rail Freight & Logistics during the course did you enjoy best?	
1	i. Rail freight interchanges ii. Opening freight terminal sidings iii. Innovation in rail freight vehicles	Both issues link directly with my new project
2	i. Rail freight system in the UK and the development ii. Rail interface	
3	i. Rail industry in the UK ii. Software designed to improve rail operations	Possibility to apply them to Thailand railway system
4	i. The vision for the rail freight system ii. Urban freight by rail	Used problem-solving skill to answer relevant questions. Made us think out of the box
5	The vision for the rail freight system of the future	Creative and out of the box thinking to generate ideas without considering reality. Can increase the participation of group members
6	Group discussion—European Commission	Fun to exchange ideas through what we read

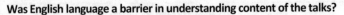

Was English language a barrier in understanding content of the talks?

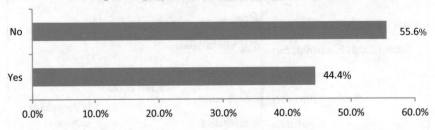

Fig. 5 Language barrier

Was English language a barrier in understanding content of the talks? (Reason for Yes)

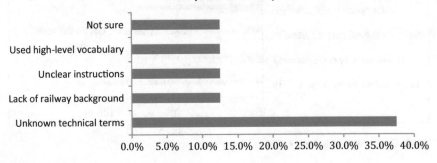

Fig. 6 Reasons for the language barrier

Would you recommend this Rail Freight and Logistics course to other delegates?

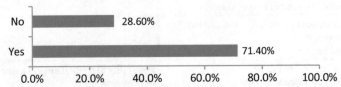

Fig. 7 Percentage of participants voting Yes or No

voted No. Figure 7 shows the graphical percentage distribution of participants voting Yes or a No.

Some of the common reasons for voting Yes were the quality of content in the talks, the participants could learn about the recent innovations in rail freight and logistics. Common reasons for voting *No* were the difficulties in understanding the English language and lack of railway knowledge by participants.

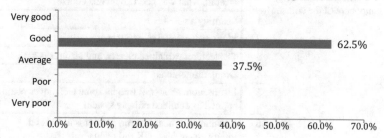

Fig. 8 Participants rating on the materials provided

Question Six: Rate materials received

Here participants were asked to rate the learning materials received during the course. The rating ranged from Very poor to Very good. The majority of the participants rated the materials as Good—62.5% while the rest rated the materials as Average—37.5%. Figure 8 shows the graphical percentage distribution of participants rating the materials provided. Specific improvements suggested by the participants are discussed later on in question 10. It is a common practice for training courses in Thailand for the participants to receive the learning material, lectures included, before the course start date. It was argued that such a practice does not always lead to a positive learning outcome and therefore it was changed. Learning material was distributed after the lectures and for each group exercise. As the participants were not accustomed to such a practice, they thought, the learning material would be of no significant use after the lectures have been given.

Question Seven: Best feature of the course

Participants were asked what they felt was the best part of the training course. This was a descriptive question. Most participants liked the fact that they were learning something new in rail freight and logistics. The lectures gave an excellent overview of how the rail freight system operates within the scope of international logistics. The group also enjoyed learning about techniques applicable to the current situation with Thai railways and rail management in the country. Table 2 shows the answers given by every participant. All participants responded to this question.

Question Eight: Worst feature of the course

Participants were asked what they felt was the worst part of this course. Similar to question 8, this too is a descriptive question. According to the feedback forms, they felt that they could not completely make use of the course due to the language barrier, lack of rail freight and logistics knowledge. Apart from this, they feel the lecturers did not explain concepts using simple words.

Moreover, their instructions during the group exercises were unclear. Table 3 shows the answers by each participant. No response from one participant.

Table 2 Participants' most liked component of the course

	Overall, what was good about this course?
1	Learning a new field
2	More pictures and clear concept
3	The lecturer explained clearly, and lecturer is kind
4	Group discussions
5	Learnt more about rail freight, some techniques could be applied to Thailand railway system
6	Learning rail management, problems in EU and improvement plan, UK rail regulation
7	Overall picture and process to operate the rail system
8	The theory is good and More knowledge about transportation

Question Nine: Suggested improvements

Here participants were asked their opinion on how this training course could be improved. Options were given, and they had to choose either Yes or No. Furthermore, they suggested methods which were not included in the options provided. Most participants felt the organisers of the course could include more online material, lectures, and group discussions, besides organising workshops. They suggested to avoid promotional materials, include technical visits and give time for research activities. Apart from this, the participants would like to see more pictures/videos while learning about new concepts and mechanisms of equipment. A glossary sheet including key terms and definitions is also suggested by the participants. A list of abbreviations should also be developed and distributed before the course start date. Both the glossary and the list of abbreviations would be of significant importance for participants to familiarise themselves with the technical terms and the jargon used during the course. This would help participants overcame any barriers to technical language, engage more easily and benefit from all learning activates involved. Figure 9 shows the graphical percentage distribution of options they would like to be implemented in this course.

Question Ten: Influence on career plans

In the last question, participants were asked if this course influenced their career paths. It was a Yes/No question; they had to give reasons to support their answer. 50% of participants felt it would influence their career plans, while the remaining 50% felt it would not influence their career plans. Figure 10 shows the graphical percentage distribution of participants voting Yes or No.

Common reasons were given by participants for Yes: career path involves the railway industry (37.5%) and believes Thailand is improving the railway system (12.5%). Common reasons for voting No: career path does not involve the railway industry (25%) and feels Thailand uses fewer railways for import and export of goods

Table 3 Participants' least liked part of the course

	Overall, what was bad about this course?
1	Cannot participate all course
2	Too many texts
3	Lengthy course and need to do work together
4	Basic knowledge not provided
5	No proper introduction to rail freight (difficult for participants with little knowledge to understand)
6	Presentation slides hard to understand (use simple words to explain), lack of pictures/video clips used while explaining
7	Need more clarity with examples, videos involving movement (e.g. hump, terminal, interchange)
8	Lack of simplified explanations with videos for proper understanding

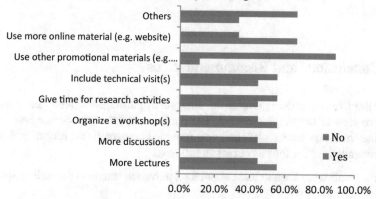

Fig. 9 Percentage distribution of preferable methods to use for improvement

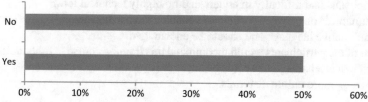

Fig. 10 Percentage of participants voting a Yes or No

Do you think this course will influence your future career plans? (Reason)

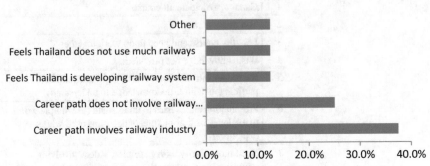

Fig. 11 Percentage distribution of reasons for voting Yes or No

(12.5%). Figure 11 shows the graphical percentage of reasons provided for voting a Yes or a No.

5 Conclusions and Recommendations

After the five-day training course, an analysis of participants' feedback showed a positive view of the training course. The participants found the course helpful and meeting their expectations. Participants provided constructive criticism and useful recommendations for future courses as follows:

- Participants were keen to learn about the innovations made in the Rail freight and Logistics industry.
- Active participation was observed in most group exercises. Participants enjoyed the exercises involving the vision for future rail freight system.
- Language barriers of participants restricted them from having a better experience during the course.
- Participants had difficulty in understanding highly technical terms.
- Communication between participants enabled them to understand course concepts in their native language and should be encouraged.
- Most of the participants would recommend this training course to other delegates.
- Participants felt this course could be improved by adding more lectures, discussions and using more online materials.

Acknowledgements The authors would like to thank Dr Somnuk Ngamchai, Ms Duangporn Teer-apabpaisitit, Ms Pakkapatee Luanpaisanon and their team in Thailand for the involvement and kind assistance for a smooth running of this course.

References

Dawson A, Braz L, Blauensteiner B, Isler C, Dias A, Asaff Y, Marinov M (2017) Evaluation of a rail-orientated researcher links workshop. In: Rail Exchange, vol 2, 2017, Sustainable Rail Transport. Springer

Fraszczyk A, Marinov M, Woroniuk C. (2012) Innovative forms of education in rail freight and logistics. In: 4th International scientific conference on transport problems, 2012, Politechnika Slaska, Katowice-Slemien, Poland

Fraszczyk A, Dungworth J, Marinov M (2015) Analysis of benefits to young rail enthusiasts of participating in extracurricular academic activities. Soc Sci 4(4):967–986

Fraszczyk A, Dungworth J, Marinov M. (2015a) An evaluation of a successful structure and organisation of an intensive programme in rail and logistics. In: The 3rd UIC world congress on rail training, 2015, Lisbon, Portugal

Fraszczyk A, Drobisher D, Marinov M (2016) Statistical analyses of motivations to participate in a rail-focused extra-curricular activity and its short terms personal impacts. In: 7th International conference on operations and supply chain management, Phuket, 2016. The Laboratory of Logistics and Supply Chain Management, Phuket, Thailand

Fraszczyk A, Weerawat W, and Marinov M (2017), Exchange of higher education teaching and learning practices between the United Kingdom and Thailand: a case study of RailExchange courses. In: Sustainable rail transport, pp 241–256. https://doi.org/10.1007/978-3-319-78544-8_13

Indian Railways (2017) Annual report, 2016

Lautala P, Edwards R, Rosario M, Pachl J, Marinov M. (2011) Universities in Europe and the united states collaborate to develop future railway engineers. In: WCRR - the 9th world congress on railway research, 2011, Lille, France

Marinov M, Fraszczyk A (2013a) Workshop on rail education: academia, industry, research, engagement, promotion, co-operation. In: 4th World conference on learning teaching and educational leadership (WCLTA-2013)

Marinov M, Fraszczyk A (2013b) Valorisation of an MSc programme. In: 4th World conference on learning teaching and educational leadership (WCLTA-2013)

Marinov M, Fraszczyk A (2014) Curriculum development and design for university programmes 301 in rail freight and logistics. Procedia Soc Behav Sci 141:1166–1170

Marinov M, Ricci S (2012) Organization and management of an innovative intensive programme in rail logistics. Procedia Soc Behav J 46:4813–4816

Marinov M, Viegas J (2011) Tactical management of rail freight transportation services: evaluation of yard performance. Transp Plan Technol 34(4):363–387

Marinov M, Zunder T, Islam D (2010a) Concepts, models and methods for rail freight and logistics performances: an inception paper. In: Proceedings media of the 12th world conference on transport research, 2010. World Conference on Transport Research, Lisbon, Portugal

Marinov M, Zunder T, Schlingensiepen J, Ricci S, Karagyozov K, Razmov T, Dzhaleva-Chonkova A (2010b) Innovative concepts for knowledge exchange, mobility and expertise in rail freight and logistics. In: Highlight 2010: sustainable development in logistics (ICLEEE) international conference of logistics, economics and environmental engineering, 2010, Vocational College of Traffic and Transport, Maribor, Slovenia

Marinov M, Pachl J, Lautala P, Macario R, Reis V, Edwards R (2011a) Policy-oriented measures for tuning and intensifying rail higher education on both sides of the Atlantic. In: 4th International seminar on railway operations modelling and analysis (IAROR), 2011. International Association of Railway Operations Research, Rome, Italy

Marinov M, Mortimer P, Zunder T, Islam D (2011b) Short haul rail freight services. RELIT - Revista de Literatura dos Transportes 5(1):136–153

Marinov M, Fraszczyk A, Zunder T, Rizzetto L, Ricci S, Todorova M, Dzhaleva A, Karagyozov K, Trendafilov Z, Schlingensiepen J (2013) A supply-demand study of practice in rail logistics higher education. J Transp Lit 7(2):338–351

Office of the National Economic and Social Development Board (2016a) Macroeconomic indicators
Office of the National Economic and Social Development Board (2016b) Thailand logistics report
Office of Transport and Traffic Policy and Planning (2017a) Action Plan for Transportation Development in Thailand
Office of Transport and Traffic Policy and Planning (2017b) Action Plan for Transportation Development in Thailand
Tsykhmistro S, Cheptsov M, Cheklov V, Marinov M (2014) Euro-Asian co-operation in rail education and research. Transp Prob: Int Sci J 9(1):103–110
Woroniuk C, Marinov M, Zunder T, Mortimer P (2013) Time series analysis of rail freight services by the private sector in Europe. Transp Policy 25:81–93

E-Sources

SRT (2018) Retrieved from http://www.railway.co.th. Accessed on 22nd Aug 2018
ThaiRailTech (2017) Retrieved from www.thairailtech.or.th. Accessed 22nd Aug 2018
SET (2017) Retrieved from www.set.or.th. Accessed on 22nd Aug 2018

Short Communication Paper

TopHat® Top-Lifting Semi-trailer—A Game Changer for Inter-Modal Freight in Europe

Phil Mortimer, Truck Train Developments Ltd

Corresponding author: pmtrucktrain@tiscali.co.uk

This discussion sets out the case for a major change in the way rail accommodates high value, time sensitive freight for national domestic and cross border traffic within Europe. Rail's share of the high value time sensitive commodity sector has been completely outperformed by the road transport sector on cost, product and service and become the accepted standard requirement for this category of traffic. Rail has not addressed the imperatives of the market and continued to press the case for orthodox inter-modal options when shippers and forwarders increasingly want to use tri-axle semi-trailers as the preferred cargo module. The TopHat concept addresses these issues. It makes the case for more rail capable trailers. This has benefits for the shippers, receivers, truckers, terminals and rail operators as well as generating real benefits in the form of reduced long and medium haul road freight, reduced emissions, better use of transport assets and reduced inter-urban traffic congestion.

Despite decades of railway reforms, changes in governance, privatization, strategies, policies and a whole lot more, rail's share of the European domestic and cross border freight market continues to decline. Road transport has won markets from rail or developed wholly new markets based around considerations of cost, agility, availability and reliability. These are the stark facts which suggest there is much more to be done if rail freight is to come back into contention for high value time sensitive (HVTS) traffic flows in any meaningful way. Road transport technologies and operational methods have found greater levels of support from shippers and the wider freight industry by offering products and services which confer upon the users, real tangible benefits in terms of cost savings, time savings and the ability to adapt quickly to evolving logistics systems and methods. Rail by contrast has largely retained a less adventurous and more risk averse supply side position but has largely failed to attract and retain traffic, particularly HVTS where its share is very small. According to EuroStat 2017, the overall European position varies widely with limited rail freight market participation in Iberia (<5%) up to 17% in Germany.

Rail has secured a more significant market position in relation to inter-modal traffic, particularly international maritime ISO container and specific flows of commodities moving either domestically or on cross border flows within Europe. The use of containers and swap bodies has not offset the loss of traffic from traditional siding to siding flows or acted as ready inter-modal option for shippers. Containers are not always ideal due to volumetric limitations and the inherent need for intermediate handling at terminals which imposes cost and time penalties. The use of containers for domestic traffic has been heavily supported by the EC but despite this their use has not been as widely adopted as proposed. The transport market prefers tri-axle semi-trailers for its primary needs. If rail is to recover markets it has lost to road transport, being able to move semi-trailers quickly, efficiently and on a cost effective basis is one potential avenue of commercial, operational and technical development.

Tri-axle semi-trailers are the workhorses of the European freight sector with an estimate million plus trailers in operation in a wide variety of traffic and commodity applications. Determining the exact size of the European trailer fleet is problematic and again exposes the limitations of European and national freight statistical sources. These trailers have effectively displaced rail from market contention by either taking over the traffic on a more cost effective basis or carrying traffic for which rail is not even considered as a credible option.

About 5% of all semi-trailers in circulation in Europe are "rail capable" in that they are able to be transferred using specialized lifting equipment to move them between road and rail. (re SAIL project) These types of trailers are lifted by grapple arms which locate onto the trailer chassis for the transfer operation. Various rail vehicle types have been developed to accommodate these trailers including pocket wagons and specialized units.

Other lifting systems including slings and bespoke swing arm wagons have been developed but are effectively niche applications and have not found widespread application. They have not induced major changes in terms of inter-modal flows. The grapple arm technology set poses problems in terms of the slow loading/offloading cycle times at terminals and the damage the arms inflict on trailer chassis, bodies and cargo contents. The loss of earning power whilst trailers are repaired plus the cost of damage to cargo has not been identified in detail again because of a lack of relevant and timely available European wide statistics and commercial confidentiality in relation to claims made. A similar position applies in North America where cargo and asset damage claims are massive.

Despite these limitations there is evidence that shippers, forwarders and wider cargo interests prefer to use tri-axle semi-trailers as the preferred cargo module for domestic and cross border traffic rather than containers, certainly ISO units for HVTS commodities. The reasoning behind this is complex and includes issues of asset ownership, availability cargo priorities, cargo characteristics and special needs for stowage and handling and the inherent "fully wheeled" position using semi-trailers which deletes the requirement for chassis pools at the load and arrival terminals.

The EC Europe Commission has expressed aspirations to achieve a significant positive modal shift to rail (30% participation in all traffic over 300 km by 2030, 50% by 2050) but has not suggested any specific operational, technical or commercial initiatives about how to achieve this. There is a growing market for high value time sensitive cargo in Europe and this is mirrored in the growth of road transport to service traffic moving under demanding imperatives.

Recent project analysis has indicated terminal and haulage costs amount to over 50% of the total cost of an inter-modal transit. Reducing these costs and making the transfer operation quicker by deleting the need for chassis pools at terminals together with enhanced safety in transfer activities are the aspirations the concept has been designed to address. By doing this an inter-modal option for shippers and cargo interests which prefer to use semi-trailers rather than containers becomes available. The rail sector has to accept that this is a market requirement and not to continue to deny this. Supply side positioning will constrain market penetration in growing markets for HVTS traffic where rail has a major potential cost, service and product advantage.

Enter TopHat®!

The TopHat® top lifting semi-trailer concept has been developed in response to the range of issues and more set out in the introduction (Fig. 1). The key question was how to make semi-trailers, which are the shippers' preferred module option, more easily rail capable and on a much wider level of availability. Existing transfer technologies are slow, inflict damage on the trailer assets, are dangerous in terminal locations or require very specialised equipment to perform the inter-modal transfer. The concept has Europe wide potential for long and medium transits where shippers and cargo interests have no inter-modal option at present.

Full sized semi-trailers are required to maximize cargo carrying capabilities. Suggesting that shippers acquire specialized equipment to comply with railway loading gauge infrastructure limitations is not an option. This requires trailers of 13.6 m length and 4.0 m height are deployed to be able to move easily, quickly, safely and with zero damage to the assets between road and rail on a routine basis. Longer trailer lengths may be required as an option if these ultimately become widely accepted within Europe. At present the main focus is on standard dry van and curtain sided units as these appear to form the largest proportion of the semi-trailer fleet.

The TopHat® concept has been developed to allow trailers, as the preferred cargo module, to be moved much more rapidly, safely and on a cost effective basis using existing ISO container lifting equipment in the European terminal network.

Fig. 1 Mock up illustration of a TopHat concept trailer

The core concept is that the semi-trailers are built with lightweight but immensely strong structural features such that the entire trailer (cargo module, chassis and running gear) can all be routinely lifted on the 40′ lifting points used for international maritime traffic.

The design adaptations to allow this are a combination of material selection and clever engineering to minimize the weight of the modifications compared with a standard road only semi-trailer. In terms of value engineering this is vital to give the TopHat® trailer the new versatility but without a high cost penalty. It is also essential that the designs do not impose a major increase in the trailer weight and also allow the trailer to offer the same cubic capacity as a standard unit.

The TopHat® concept addresses a real issue that is constraining modal shift to rail, particularly to service growing volumes of HVTS traffic currently moving in road only mono-modal transits. It adds capability and versatility to trailers for inter-modal applications by using a combination of widely available established orthodox technology sets in an innovative combination. The ability to top lift the trailers uses existing terminal lifting equipment is vital as is the ability to operate either in a full inter-modal or mono-modal role as dictated by operational and commercial requirements. The TopHat option is based around the central premise of using full sized trailers in conjunction with widely available standardized lifting equipment and technologies. It is based around the movement of the entire trailer including road running gear without recourse to specialized lifting arms/grapples or more complex sling technologies. It adds inter-modal capability to trailers with a minimal cargo weight or volumetric penalty as well as giving faster, safer and more cost effective loading compared with existing trailer lifting systems.

The concept is simple but opens up options to move trailers by rail and secure a sustained modal shift with the attendant benefits in terms of emissions reduction, enhanced safety, reduced infrastructure attrition and congestion. It potentially provides a robust, practical and simple inter-modal solution for some current road based freight particularly that which is mandated to be moved in trailers. Much of this traffic is currently beyond rail's generic capabilities and commercial reach either because of the lack of inter-modal services or cost efficient means of routinely moving trailers quickly and reliably between road and rail.

The TopHat solution potential potentially enhances shipper experience by adding significantly to shipper/forwarder choices for inter-modal freight for European domestic and cross border traffic applications. It adds to existing inter-modal terminals' portfolio of capabilities. The new concept could underpin of both the development of new inter-modal services and add new capability onto existing inter-modal offers by allowing a bigger mix of trailers and containers to be carried on trains.

The freight market in Europe is subject to many influences including globalization, increased levels of urbanization, concerns over energy efficiency and emissions. Cargo characteristics have changed in response to a move from basic heavy industrial structures to those based upon lighter but more valuable characteristics. Volume rather than weight has become a key market issue for shippers and transport service providers. This market is growing as heavy industry declines (ref DG Move) and rail has to respond to this or become a marginal player. Participation in the demanding high value time sensitive sector demands a very different approach in terms of agility, responsiveness, availability, consistency and reliability as basic service tenets. The tri-axle semi-trailer component of intra-European traffic is growing. The TopHat® concept is to build on this using the widely recognized work horse with minimal modifications at the time of manufacture to yield an inter-modal capability without com promising road operations. Existing trailer on train technologies require excessive specialization at terminals and this acts as a real limitation on rail's market penetration. TopHat® massively widens the portfolio of inter-modal and co-modal capabilities and opens up technical, operational and commercial options that are not feasible at present.

The TopHat® concept is primarily aimed at new build trailers with the modifications being an integral part of the manufacturing activity. The initial applications are planned for dry vans and curtain sided trailers as these form the largest component of the European trailer fleet. Other applications including tankers and open topped trailers could also be developed. Retro-fitting existing trailers is an option but would need to be cognizant of the trailer age and technical condition at the time any such retro-fit proposal.

A major benefit flowing from the TopHat® concept is the potentially longer trailer life as a direct result of the top lifting option. The grapple arm system inflicts damage to trailer chassis, cargo bodies and contents. Trailers used in this method are normally limited to 5–7 years active service life. The TopHat® option should build significantly on this, possibly doubling the current life expectancy. This should appeal to owners, operators and leasing companies.

The TopHat® concept is being pursued through a funded development project with the application off the prototype trailers into a range of European domestic and cross border commercial traffic applications with key shippers, forwarders and trucking companies. The intention is to secure buy in and first mover interest and also to identify any detailed design or engineering issues arising from real traffic applications. This should also reinforce the investment case for users, owners and operators. It is potentially a major catalyst for change in the inter-modal freight market and could seriously reinforce rail's longer term market aspirations.

Printed in the United States
By Bookmasters